45⁰⁰

D0897663

Deming's 14 Points Applied to Services

QUALITY AND RELIABILITY

A Series Edited by

Edward G. Schilling

Center for Quality and Applied Statistics
Rochester Institute of Technology
Rochester, New York

Deming's 14 Points
Applied to Services

A. C. ROSANDER

Marcel Dekker, Inc.
New York • Basel • Hong Kong

ASQC Quality Press
Milwaukee

Library of Congress Cataloging-in-Publication Data
Rosander, A. C. (Arlyn Custer)
   Deming's 14 points applied to services / A. C. Rosander.
     p.   cm. — (Quality and reliability ; 24)
   Includes bibliographical references and index.
   ISBN 0-8247-8517-7 (alk. paper)
   1. Service industries—Quality control. 2. Deming, W. Edwards
(William Edwards) I. Title. II. Title: Deming's
fourteen points applied to services. III. Series
HD9980.5.R674 1991
658.5'62—dc20   91-6799
                                         CIP

This book is printed on acid-free paper.

Marcel Dekker, Inc.
270 Madison Avenue, New York, New York 10016

ASQC Quality Press
310 West Wisconsin Avenue
Milwaukee, Wisconsin 53203

Current printing (last digit):

10 9 8 7 6 5 4 3 2 1

PRINTED IN THE UNITED STATES OF AMERICA

To Dr. W. Edwards Deming, who saw the need for a new type of management based on statistics, and was required to attain continuous quality improvement; and to Jane Choate, my skilled and reliable secretary, for error-free performance, I am very grateful.

About the Series

The genesis of modern methods of quality and reliability will be found in a simple memo dated May 16, 1924, in which Walter A. Shewhart proposed the control chart for the analysis of inspection data. This led to a broadening of the concept of inspection from emphasis on detection and correction of defective material to control of quality through analysis and prevention of quality problems. Subsequent concern for product performance in the hands of the user stimulated development of the systems and techniques of reliability. Emphasis on the consumer as the ultimate judge of quality serves as the catalyst to bring about the integration of the methodology of quality with that of reliability. Thus, the innovations that came out of the control chart spawned a philosophy of control of quality and reliability that has come to include not only the methodology of the statistical sciences and engineering, but also the use of appropriate management methods together with various motivational procedures in a concerted effort dedicated to quality improvement.

This series is intended to provide a vehicle to foster interaction of the elements of the modern approach to quality, including statistical applications, quality and reliability engineering, management, and motivational aspects. It is a forum in which the subject matter of these various areas can be brought together to allow for effective integration of appropriate techniques. This will promote the true benefit of each, which can be achieved only through their interaction. In this sense, the whole of quality and reliability is greater than the sum of its parts, as each element augments the others.

The contributors to this series have been encouraged to discuss fundamental concepts as well as methodology, technology, and procedures at the leading edge of the discipline. Thus, new concepts are placed in proper perspective in these evolving disciplines. The series is intended for those in manufacturing, engineering, and marketing and management, as well as the consuming public, all of whom have an interest and stake in the improvement and maintenance of quality and reliability in the products and services that are the lifeblood of the economic system.

The modern approach to quality and reliability concerns excellence: excellence when the product is designed, excellence when the product is made, excellence as the product is used, and excellence throughout its lifetime. But excellence does not result without effort, and products and services of superior quality and reliability require an appropriate combination of statistical, engineering, management, and motivational effort. This effort can be directed for maximum benefit only in light of timely knowledge of approaches and methods that have been developed and are available in these areas of expertise. Within the volumes of this series, the reader will find the means to create, control, correct, and improve quality and reliability in ways that are cost effective, that enhance productivity, and that create a motivational atmosphere that is harmonious and constructive. It is dedicated to that end and to the readers whose study of quality and reliability will lead to greater understanding of their products, their processes, their workplaces, and themselves.

Edward G. Schilling

Preface

W. Edwards Deming revolutionized management in manufacturing and service industries by insisting that top management be responsible for continuous quality improvement.

I have already shown in *The Quest for Quality in Services** that quality of services is very different from the quality of manufactured products. Quality of services requires a different approach. For example, safety is the most important quality characteristic in services. In transportation, health services, power utilities, construction, and chemical plants, where one human error can be fatal, zero errors is a "must."

The customer does not buy from the CEO. The customer is served by salespersons, clerks, attendants, and other lower-level personnel. They determine the quality service the customer receives.

Dr. Deming's 14 points, the focus of this book, are applied to quality

*Milwaukee: ASQC Quality Press and White Plains: Quality Resources, 1989.

service. This means that several of the points have to be interpreted in terms of the realities of quality service.

Service industries buy defective products—defective medical equipment, defective drugs, defective buses, defective airplanes. The service industries must insist on evidence, such as quality-control charts, that each shipment meets their requirements. Zero defects is the goal.

Prevention of error is crucial to a service quality program. Other important non-quality characteristics are unwarranted delays, wasted time, unnecessary service, lost time, and incompetent managers and employees. Safety, the most important of all, means zero errors.

Dr. Deming's statistical approach is based on principles derived by Shewhart. This distinguishes special causes from chronic causes, a fundamental concept in statistical control. Assignable causes include both, as Shewhart shows. The role of statistics is basic.

Certain techniques helpful to services but neglected in a factory include learning curve and modern random time sampling analysis. Dr. Deming devotes one chapter in each of his books to service quality. He concentrates on examples from manufacturing but uses over 30 statistical problems from the factory.

This book begins with an overview of the development of quality control and of Dr. Deming's role. His 14 points are listed and explained in Chapter 2. The final two chapters outline a program for implementing quality-control techniques, from evaluating and training employees to setting long-term goals and company-wide policies. Like Deming, I have included an introduction to statistical techniques.

A. C. Rosander

Contents

Deming's 14 Points Applied to Services

CHAPTER 1

Introduction

Background

To understand the situation since 1950 with regard to quality control, it is necessary to understand the national economic situation, the state of statistics, and statistical quality control.

After World War II the United States was a seller's market because 100 million people had a pent-up demand for goods and services that had accumulated for five years. Production was speeded up everywhere. The emphasis was on quantity; quality was ignored.

Statistics was just beginning to emerge as a science, but sound books were still ahead. Very few had heard about Shewhartian quality control, let alone understood it. Eugene L. Grant published the book *Statistical Quality Control* in 1946. During the war, a few thousand people were trained in the rudiments of statistical quality control. They formed the American Society for Quality Control, which was founded in 1946.

The responsibility for quality control was taken away from people who did not know what steps to take to ensure quality control. The major obstacle was trying to get people interested in quality and statistical quality control. There was no reason that they should be interested in quality control.

In trying to demonstrate whether Shewhart's theories would work in the real world of manufacturing and services, we had to learn enough statistics to apply these theories to real problems. An individual can innovate, but a group has to implement. Since most of us were in middle management, this meant we had to *convince* management higher up, and *train* people lower down.

Of these difficulties, convincing management was the real problem. We discovered top management was the obstacle long before many others did, because we were operators, not consultants. The fact that we wrote statistically oriented papers did not mean we were not faced with people problems—with hiring, supervision, training, appraisal, and quality performance.

The foregoing illustrates why we couldn't make quality everybody's business until we had made it somebody's business. We sought just one high-level official to understand and accept probability sampling and quality control as superior to anything the organization had, and to give us the green light to start a project. Some of us started a project without the approval of top management.

Quality Control in Japan

From 1950 through 1975, Japan accepted quality control at top levels due to the efforts of W. Edwards Deming. We started quality control, but it was only a small beginning. Its growth led to many successful applications. Quality control never became a national movement in the United States, although we originated it.

We didn't realize that postwar prosperity had come to an end until we had to meet foreign competition in automobiles, electronics, textiles, shoes, and steel. We must meet foreign competition in today's products, which are not the inferior goods of the past, but high-quality products. Japan is now a major competitor.

The weaknesses of American business that very few American executives saw, but that key officials in Japan recognized, are the need for:

1. Short-term planning
2. Stressing quality production

3. A work force adequately trained in mathematics, physical, and biological sciences; improved education
4. More savings
5. Production rather than consumption
6. A better work ethic
7. Cooperative rather than adversarial factory production relationships

Deming's Influence in the United States

Dr. Deming's contributions to quality didn't receive much attention until the 1970s and 1980s. His 14 points calling for a revolution in American management were first published in 1982 in his book *Quality, Productivity, and Competitive Position. Out of the Crisis* was a 1986 revision of this book. The goal was continuous quality improvement based on statistical techniques.

The 1982 book assumes the reader knows certain statistical principles, such as five applications of the binomial count and the binomial proportion, five applications of the Poisson distribution, and $\bar{X}$ and R charts. These are explained in Chapter 3, Section 7, "Introduce Statistical Techniques."

Deming's appearance in the television program "Japan Can Do It, Why Can't We?" and his series of seminars brought him national attention. His Japanese connection paid off when the United States had to face Japanese competition. Deming knew the secret. American management did not. The issue was how to build quality into products and services. We concentrate on the latter.

Scope of Quality of Service

Service is a face-to-face situation in which the customer deals directly with a salesperson or clerk. Quality of service is what these salespersons or clerks give us. The quality of services does not come from the CEOs. Unless high-level management understands that quality improvement is a continuous program, and convinces all lower-level workers to adopt the same view, salespersons and clerks will not give us quality service.

Quality of service is also associated with professional and technical workers, who determine the quality of service the customer receives. These workers range from physicians to auto repair technicians.

A physician diagnoses an illness. Quality service means a correct diagnosis and an effective treatment to free the patient of illness. An auto repair technician must also make a correct diagnosis and eliminate the problem's cause. This is quality performance—getting rid of trouble.

It is prudent to remember that the customer does not buy from the CEO.

Definition of Quality Service

All service organizations depend on a steady flow of buying customers for their success. This does not necessarily mean that they have to be customer-oriented or customer-driven.

Quality means meeting the requirements of the customer. It is based on prevention of non-quality characteristics—errors, defects in purchased products, wasted time, delays, failures, unsafe working conditions, unnecessary service, and unsafe products. All service should be aimed at meeting the customer's requirements by eliminating these non-quality traits.

All employees, from the chief executive officer to the lowest-paid worker, should actively engage in meeting these requirements, as well as being alert for new non-quality traits.

Examples of customer requirements are:

Water: 100% reliability
Gas: 100% reliability
Electricity: 100% reliability
Telephone: 100% reliability
Clothing and shoes: must fit
Medicine: effective, no side effects, reasonably priced
Medical service: correct diagnosis, effective treatment
Checking account: no errors
Mail: no delay
Retail store: pleasant, helpful salesperson; honesty

Multiplicity of Quality Characteristics

Customers purchase services and goods with a *multiplicity* of quality characteristics, due to the very large number of products and services they buy. This large number results in quality that is *heterogeneous* because the quality characteristics desired are *widely* different depending on whether one is evaluating, for example, hospital services, airline transportation, shoes, hotel rooms, insurance policies, automobiles, computers, lawn mowers, or banking services. These are just a few examples of the *variability* of quality among products. The customer is concerned not with the quality of only one product or service but with thousands of products and services.

New and Discontinued Products

The customer is aware of new products (e.g., foods and drugs) of questionable value that appear almost weekly in the supermarket. On the other hand, several highly desirable products have been discontinued. Examples of acceptable-quality goods that some stores no longer carry are:

- Imported clothes and shoes
- Petite women's clothes
- Sodiphene (antiseptic)
- Saffola margarine
- Unguentine (salve for burns)
- Tincture of iodine (antiseptic)
- Men's summer clothes
- Pyrex
- Cast-iron skillets

This practice ignores what many customers want. Stores cater to market pressures that favor a profitable group, e.g., middle sizes in clothing and shoes. Or they try to sell the monstrosity "One size fits all."

A Sample of One

The customer buys a sample of one. There is no frequency distribution, no mean, and no variance. Only after repeated buying of the same product or service over months or years does the customer get some idea of quality.

These commonly repeated purchases include carrots, apples, bread, automobiles, home insurance, health insurance, automobile insurance, and telephone service, as well as services of physicians, dentists, and lawyers, to mention just a few.

Service Management and the Customer

Service is the latest interest of management, especially in manufacturing. Consumer-driven quality comes from the factory, but a few service organizations have learned the secret of the satisfied customer.

In the factory, management is beginning to bring other factory customers into a closer relationship with their operations. This is peculiar to manufacturing. Service agencies use advertising more than customer contacts; customer contacts are largely ignored.

Management must take the lead in promoting and implementing customer-driven quality in the service industries. This means the customer is given top consideration in the customer–management–service cycle.

Prevention of Non-Quality

Quality consists of meeting the customer's requirements. When these requirements are not met, non-quality characteristics exist.

Our purpose is to describe these non-quality characteristics of services and the steps management has to take to prevent them. When management eliminates them, customer requirements will be met, costs will be reduced, and productivity will be increased.

The Deming Plan

Deming's plan is to show top-level management how to manage. This applies to service industries as well as to manufacturing.

Management means mastery of production, supervision, and training—all aimed at building quality into a product or service the customer will buy. The plan calls for a continuous program of quality improvement.

We explain each of Deming's 14 points as applied to service industries. This is followed by a detailed description of how to implement a service quality program in practice—its characteristics, their occurrence, their analysis, and the prevention of non-quality.

CHAPTER 2

Deming's 14 Points

Deming's 14 points are based on Shewhart's theories. Although Shewhart's techniques are the key to quality, or the elimination of non-quality, Deming goes much further.

The 14 points offer no exotic theory of behavior. There is no mention of theories X, Y, or Z. The 14 points are pragmatic; they are based on actual observations of what goes on in a factory or service organization. Deming's theories are derived from first-hand observation—hence the realities of his insights. In every example he selects, there is a lesson to be learned.

The Deming–Shewhart plan contains much that originated with Shewhart, such as:

- "Assignable causes" (special or peculiar causes), to be corrected by employees.
- "System causes," which only high-level officials can solve.
- Cooperation between marketing, design, engineering, manufacturing, and other departments.

- Division of suspect data into "rational subgroups" for best control—when three groups accumulate into one group, control charts on each group are needed to measure differences in variation.
- Quality control to save rework and cut costs.
- Quality techniques and charts originated by Shewhart.
- Controlled conditions.

Deming, looking at the larger picture, saw that management must accept leadership if a quality program is to be effective.

1.   Create Constancy of Purpose for Improvement of Product and Service

Constancy of purpose means a steady enduring, a never-ending goal. Improvement results in continuous elimination of non-quality characteristics: errors, delays, wasted time, objectionable behavior and attitudes, defects in purchased products, failures, unsafe working conditions, and unnecessary service. These characteristics are uncovered by problem analysis, audits, customer complaints and ideas, and employee suggestions.

For service industries, constant purpose aimed at improvement includes the following:

- Long-range plans
- Expansion to handle more customers
- Progressive reduction of the number of non-quality characteristics with zero as the goal
- Continuous training
- Communication from top to bottom and vice versa

Specific Activities

A new constant purpose requires specific activities in a continuous quality-improvement program:

- New ideas
- New products that sell and meet competition
- New products that are designed to enable a company to stay in business
- New and better methods, e.g., probability sampling

- New and improved processes
- Better equipment and machinery
- Better-trained workers
- Better-trained managers
- New services to satisfy the customers' needs and demands; improvement in existing services
- Improved maintenance of machinery and equipment
- Improved office maintenance
- Better safety program
- Continuous error-prevention program
- Continuous customer surveys
- Reduced error rates
- Reduced wasted time
- Reduced delays
- Improved quality behavior and attitudes of all personnel
- Immediate attention to the customer

Quality Planning

Constancy of purpose calls for applying quality improvement to *current* situations and problems. It means applying quality improvements methods *immediately*—e.g., to billing errors.

It requires planning to attack situations that arise over time, such as new services, new hires, and continued training.

It requires statistical capability to apply the simple techniques used by Deming. This means:

- Defining a quality policy
- Setting quality standards
- Defining key quality concepts operationally
- Exerting constant leadership in all aspects of quality
- Maintaining an experienced staff; rewarding workers to help prevent job hopping
- Stressing that the customer's needs and preferences constitute the heart of any quality program
- Providing means of reaching the above goals

Management Turnover

A company cannot have a constant purpose if top-level managers become transients who shift from one organization to another. A study of top appointed federal officials showed that they stayed in office only 18 months—hardly time to develop a constant purpose. The federal register of civil service executives maintained in Washington actually promotes mobility.

An excellent example of the loss this practice causes is the case of a tax administrator. His many years of tax experience rated him an expert administrator. The President selected him to head another federal agency that was having trouble getting started, an agency he knew absolutely nothing about. The fallacy was that any good administrator could save this new agency. The result? A fine tax administrator was wasted.

Weaknesses of Management

Typical weaknesses of management include:

- Aversion to change
- Mobility—the transient manager
- Lack of relevant knowledge
- A negative attitude
- Putting technical decisions in the hands of those not competent to make them
- Favoritism
- Relying on salesmanship, not quality
- Overriding engineers' decisions
- Meeting competition with anti-quality performance: poor maintenance, poor service, unreliability, or faulty equipment
- Short-range financial policies and goals
- Indifference to quality improvement
- Inability to lead
- Lack of communication with lower-level managers and workers

Obstacles

Management may want a quick fix: The quality control department, a statistician, or a quality council will be responsible for quality. In such a case, management approves of quality only if it is assigned to someone else, while management concentrates on profits, balance sheets, and other financial matters. Instead, management must accept continuous responsibility to lead in quality performance.

Finance as Major Interest

Some businesses shortchange production to concentrate on financial matters:

- Profit and loss
- Interest and dividends
- Assets and liabilities
- Bonds
- The stock market
- Rate of return
- Sales
- Market share
- Gross and net income
- Cost and income accounting
- Loans
- Budget
- Buyouts

When these predominate, management has little or no interest in quality improvement of products or services.

Basis for Change

All operations do not contain information that will lead to improvement. An example is where probability sampling leads to a continuous improvement or a breakthrough. Examples are:

1. The Bureau of the Census originated the nationwide probability sample of 60,000 households that gives current estimates of employment and unemployment.
2. World War II required quarterly estimates of critical materials. The orthodox way of estimating for a future quarter was to add 15 percent to the previous quarter.

   Probability sampling of companies changed all this. A randomly selected, highly stratified sample with ratio estimation gave a much better and quicker set of estimates. The 15-percent factor didn't suggest use of a stratified random sample. It was pure guesswork.
3. The audit control sample of the Internal Revenue Service measured the magnitude of noncompliance on Form 1040. Prior to this survey, the incidence of noncompliance was only wild guesswork.

4. Random time sampling (RTS) using the minute model produces frequency and cost of various work activities. It decomposes joint costs. It shows how workers are using work time. Current activities do not use RTS as an analytical and improvement method as they should.

2. Bring About a New Age

A proclamation of a new age of acceptable-quality products and services is fine, but the real question is: How do you bring it about? This question applies not only to the economic system, which doesn't want to meet competition—foreign or domestic—but also to the educational system, from the elementary school through the graduate school.

With about 25 million illiterates in this country and low intellectual standards in the schools and colleges, we are facing a critical situation. The schools need to stress mastery of important subject matter and basic knowledge; quality performance in all schoolwork, including mathematics; the ability to think, which requires application of computational, vocabulary, written, and speaking skills to knowledge; sensitivity to accuracy; and alertness to problems. Critical thinking is a necessity.

The foundations of quality performance in the working world will have to be established and developed in the schools and in the home. We should make very clear that these measures are necessary if this country is going to survive. This is not being done now. Companies should not become elementary or high schools, spending $25 billion annually to teach basic knowledge and skills.

We need to do more. Companies need to develop programs for *error prevention* in every activity and operation. Likewise, they need to develop programs to prevent *wasted time* in every activity and operation. They need to develop programs centered on *customers'* desires, preferences, complaints, and ideas for improvements. These programs must be continuous and have top priority. Without these steps, a new age is only a dream.

Presumably, the major objectives of Deming's points 6 (training on the job), 7 (institute modern methods of supervision of production workers), and 13 (vigorous program of education and retraining), are to create better products and services through prevention of errors and wasted time; establish reliable and humane services; and improve workmanship and performance. All these will continuously reflect attitudes of accuracy, care, and concern. Specific plans, programs, and actions are required. The customer will be the center of attention, not an outsider.

A new age is not around the corner. We can no longer live with the following in services:

- Errors
- Delays of all kinds
- Wasted time of all kinds
- Defective products
- Non-quality behavior
- Non-quality attitudes
- Unsafe methods and actions
- Failures

Examples of Reduction of Defects and Scrap

The following two examples show reduction of defects and expense of scrap over a three-year period.

These improvements were realized in a computer company's integrated circuit testing facility from 1986 through 1988.[1] Defects were reduced from 1000 to 270 per year—a 73-percent reduction. This meant no rework on 730 units, with a huge reduction in costs and a big increase in productivity (Figure 2.1).

The error rate declined in fractional form:

$$\text{from } \frac{1.00}{1000} \text{ to } \frac{27}{100,000} = \frac{0.27}{1000}$$

This shows the 73-percent decrease in number of defects: $1.00 - 0.27 = 0.73$. The decline over three years showed that the company had a continuous quality-improvement program in place.

Figure 2.2 shows the corresponding money savings in cost of scrap (defects). Over three years, the cost of scrap went from $600,000 to $74,000, a decrease of $526,000, or 88 percent.

This savings required no capital investment, no new equipment, no new machinery, and no new laboratory apparatus—just know-how. As we have known for years, all this improvement requires is knowledge, particularly statistical knowledge.

The cost of quality is the cost of nonconformance as measured by the number of defects. The cost of non-quality in this case would have been $526,000 over a three-year period.

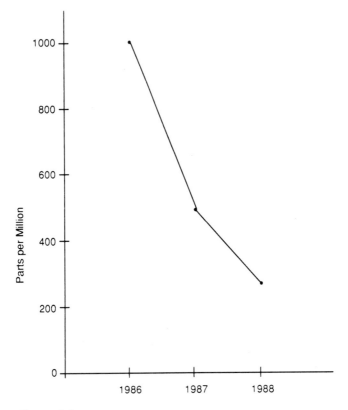

Figure 2.1
Number of defects, 1986–1988, in parts per million. (From Ref. 1.)

This example shows the tremendous gains that can result from a quality-control program based on the right kind of knowledge. This company existed in an age dominated by a constant striving for quality improvement in products and services.

3.   Avoid Massive Inspection

One can avoid massive inspection in a number of ways:

1. Contract with only a few vendors.
2. Buy from a few high-quality producers.

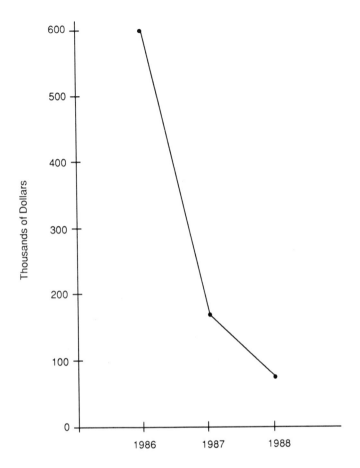

Figure 2.2
Cost of scrap, 1986–1988. (From Ref. 1.)

3. Cooperate closely with vendors.
4. Establish mutual exchange of problems and quality standards to meet operating requirements.
5. Arrange vendee–vendor agreement or contract to give vendee only what operates successfully.
6. Arrange just-in-time delivery.

Table 2.1 provides additional examples.

Table 2.1
Examples of Malfunctioning

Event	Corrective Action
Mixup of drugs, gases, and medicine in hospitals and nursing homes (lookalike containers)	• Use different storage places • Use different shapes and sizes of containers • Use large identification labels • Monitor gas suppliers
Transit company orders 100 buses, but 50 are defective	• Return defectives to the factory to correct defects (the transit company repaired them until they had "a good average bus")
Defective household water meters—faulty electric circuit connected to meter fastened to house	• Redesign meter • Install waterproof material between terminals
Pacemakers with defective electrical parts made by foreign and domestic manufacturers	• Redesign to correct faults

Reducing Inspections and Rework

1.   At Deere and Company, manufacturers of agricultural machinery, substantial savings resulted from the introduction of statistical quality control methods into machine operations. Charles Wiman, president of the company, wrote:[2]

> At our Waterloo plant, for example, we have a control chart on over eighty-five percent of our machines. Before these control charts were placed before each operator, the percentage of parts that had to be scrapped or reworked because of inaccuracies ran about 5 percent. Within three years this figure has been reduced to 1.1 percent and I am confident that it will be reduced still further. Where we formerly had a large group of men in rework departments, today it is possible to assign these men to production.

Nor was this all; it was possible to reduce the amount of inspection but at the same time improve the quality of the outgoing product. Wiman's report continued:

> In one machine department before quality control we had 14 inspectors sorting the materials produced because of the high percentage of scrap and rework. After quality control and after making corrections on machines and tools, and training operators, foremen, and superintendents in this work, we reduced the number of inspectors to 6, and the percentage of rejects was reduced from about 12 percent to three percent.

The use of statistical quality control not only helps the manufacturer; it also puts the pressure on the manufacturer's suppliers to improve their quality by adopting the same methods. Wiman noted that:

> The day is not far distant when a manufacturer probably will specify to the vendor that a certified control chart must be submitted with each shipment of material. . . . The percentage of rejects of (a certain vendor) was running 16 percent but after quality control was applied (by the vendor) within 30 days the figure was reduced to 2.5 percent, and we expect still further improvements.

He also warned plant managers that inspection reports should not be filed away in a drawer with little or no corrective action being taken. The value of the statistical quality control approach is that it makes direct and immediate use of the inspection reports and inspection data to improve the production process.

Wiman observed that statistical quality control techniques can be used not only in factory production but also in such important areas as expense accounts, inventories, turnover, receivables, sales, and customer complaints.

2.   J. W. McNairy of the General Electric Company reported that in one case a sample audit of outgoing electric cordset shipments revealed that too high a percentage of defective products were getting by a battery of inspectors making a 100-percent inspection at the end of the line.[3] After a continuous sampling plan was installed, the percentage of defectives dropped to less than one-tenth of one percent. Not only was the quality of the product improved as a result of this sample inspection plan, but production increased 60 percent. In this connection McNairy made the following significant observation, which has been substantiated many times in a variety of situations:

One of the most insidious effects of 100 percent inspection of a simple elementary type product is the deadening effect on the person performing the work. Operators doing repetitive operations fail to catch defects occasionally because of fatigue, distraction, or the sheer monotony of the job. In our experience better outgoing quality is obtained in many instances, on a statistical basis, if the samples withdrawn are subjected to a completely thorough inspection, which invariably can be carried out more deliberately and under much better conditions.

3.   An example of how statistical quality control was used to improve service operations was reported by D. W. Lobsinger of United Airlines.[4] The company applied statistical quality control to its food-service operations and eliminated 87 percent of the waste in 11 months. It applied a similar technique to errors related to payload-control messages received at a central headquarters, and as a result lowered the error rate to one-fifth of what it had been five months before. This type of control was also placed on errors made in connection with passenger reservations, with good results. Finally, as a result of making a quality-control study of cabin-service complaints, the company was able to reduce the volume of such complaints by 50 percent.

4.   J. M. Ballowe of Alden's, a mail-order house in Chicago, described one of the earliest attempts to use statistical quality control in office operations.[5] The problem involved the number of errors made by clerks in filling mail orders, errors such as sending out the wrong size, color, or item. These errors were costly because they led to customer complaints and possible loss of business.

Inspectors recorded the number of errors and plotted the total error rate of all clerks daily on a very large wall chart in sight of all the workers. The very fact that data were collected on errors and exhibited on this chart, in itself, reduced the error rate from 6 percent to 1 percent. Upper control limits should not be shown on these wall charts for fear that they will be misinterpreted by the workers; however, management used these limits in interpreting the results. As a result of this experience, statistical quality control was extended to several other departments.

Summary of Results

The results of implementing quality-improvement methods at Alden's were as follows:

1. Reduced scrap and rework cut costs and improved quality.

2. Many employees in the rework department were assigned to production, thereby improving productivity.
3. The reduction in the reject rate from 12 percent to 3 percent cut costs and improved quality.
4. The number of inspectors was reduced from 14 to six, thereby improving productivity.
5. Rejects by vendors decreased from 16 percent to 2.5 percent.
6. Using quality control charts was easier and more effective than filing inspection reports.
7. Improvements were made in supportive functions.
8. Sample inspections were more efficient than mass inspection.
9. Waste was reduced 87 percent in 11 months.
10. The error rate was reduced from 6 percent to 1 percent.

4. Buy for Quality, Not for the Price Tag

The purchasing agent not only buys for quality; he also buys to *save time.* Saving time results from more efficient deliveries with the goal of just-in-time arrivals. Improving quality can be associated with a decreasing amount of time to make a finished product. This is not contradictory to buying for quality; it *is* buying for quality.

Figure 2.3 shows the number of days required to turn out a finished project. Over a period of three years, the throughput time in days was reduced from 35 the first year to 9 the second year to 3 the third year. This was an amazing 91-percent reduction in time. The result was an increase in productivity with an improvement in quality. (All five of these charts represent the performance of the same department of the same company for three consecutive years.)

Figure 2.4 represents the volume of inventory during these same three years. This resulted by arranging purchases to be delivered just in time, saving the high cost of carrying the usual inventory.

The inventory was reduced from 2000 units the first year to 288 units the second year to 120 the third year. This was a 94-percent decrease in inventory.

There are no details available on how these just-in-time deliveries were planned and implemented.

The characteristics that showed the largest decreases during the first year were:

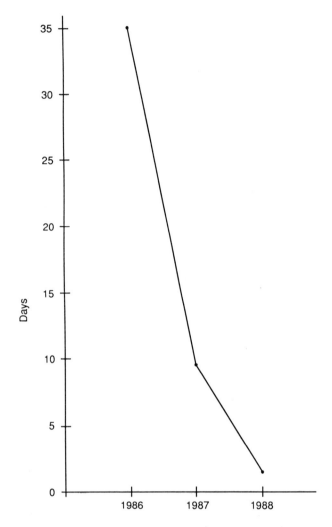

Figure 2.3
Throughput, 1986–1988. (From Ref. 1.)

1. Defects (number)
2. Cost of scrap (dollars)
3. Throughput (days)
4. Inventory (units)

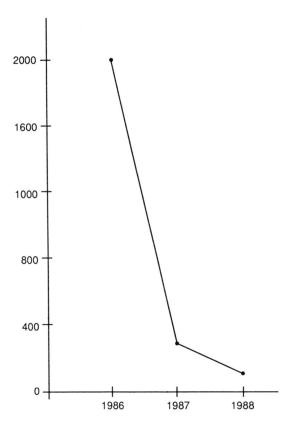

Figure 2.4
Inventory, 1986–1988, in 1000 units. (From Ref. 1.)

The decrease between the second and third years was much less, as the graphs show.

These charts show constant improvement in quality and productivity, sharply reduced costs of scrap and inventory, and an increase to 99 percent in scheduled performance (Figure 2.5).

Scheduled Performance	
1986	85 percent
1987	89 percent
1988	99 percent

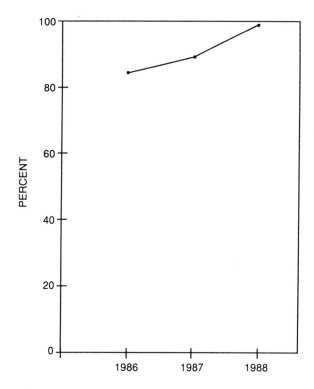

Figure 2.5
Scheduled performance, 1986–1988. (From Ref. 1.)

Example of Good Quality

I have four Hewlett-Packard computers: a 9810A with a statistical read-out memory, a 25, a 32E, and a 41C. The oldest is 12 years old and not one of them has ever needed to be repaired. This is an example of buying for quality and not the price tag (although on the first purchase, the HP trademark had no influence since the company was unknown to the author). This is a case where quality is built in. Statistical evidence shows quality is built in and that there is little or no need for any inspection.

Examples of Poor Quality

A city transit company was forced to buy buses from the lowest bidder. Among the first buses delivered, a dozen were defective. These were not returned to the factory but were repaired in the transit company's garage; a few men from the factory helped. No attempt was made to improve these buses in the factory or eliminate the causes of these defects.

A Denver schoolbus literally collapsed in the street. This is another example of buying for price and ignoring quality.

There was no attempt to ensure that quality was built into these buses in the factory. The company did not have the slightest idea that all these defects should be prevented in the factory.

Contracting Out: Examples of Buying ''Cheap''

Government contracts are let to private concerns because it is claimed that they do better work at a lower cost than government workers. These are the so-called lowest-price companies. Private concerns claim that contracting out is better and cheaper; government workers deny these claims for legitimate reasons. Consider three examples.

1. The job of cleaning a large government office building housing about 4000 employees had been performed by the General Services Administration (GSA). But then a change was made, and the job was given to a private company. We do not know the cost, but we know the quality of the work. Every time the company cleaned, furniture was scattered. Clothing racks, chairs, and wastebaskets were left in the wrong offices. "Find your furniture" became part of the office workers' routine.

This situation had not existed under the GSA. This contractor's services may have been cheap, but they lacked quality. There were so many complaints that the job was finally returned to the GSA.

As this example shows, private contractors will not necessarily do a better job, and they may cost more.

2. A federal nationwide sample study was contracted out to a small company new to the business. The study called for processing an annual project of over 150,000 documents. The value of the contract exceeded $400,000.

The job was taken out of the hands of a staff that had many years of experience with it, including its techniques and financial problems.

This was like trading a Thoroughbred for a nag. The contractor's staff consisted of a physicist, electrical engineers, and other professionals who simply did not understand the data processing of a complex sample. This was not a simple clerical job, but one that required understanding of scores of items, some of a technical nature; the detailed instructions covered about 100 pages.

The programmer made a mistake in programming the sampling error.

This mistake was carried over into publication, although the correct method was included in a report to the vice president.

The relevant procedures were used in the training process by a professional who had many years of experience on this same project. She built quality into her job; the rest did not.

The publication was issued despite the errors. Low cost and inexperience do not pay, nor do they improve quality.

3.   The U.S. Department of Commerce let a contract for a sample study to a well-known consulting firm. Sample design was not one of the firm's strong points.

After several months, the Secretary of Commerce called a meeting to discuss the progress made. He also called in two sampling specialists from two other independent agencies.

The three persons employed by the contractor described their work. It turned out that they were drawing the sample on a selected day of the month. The two sampling specialists pointed out that this was not a probability sample based on random selection.

As a result, the contract was canceled, and the job was assigned to the Bureau of the Census. A prestigious firm is not necessarily competent in solving all problems. The consulting firm was the wrong choice for this project.

5.   Continuously Hunt for Areas to Be Improved

Areas to examine for problems in services include:

- Data collecting, sampling
- Purchasing
- Personnel
- Training
- Behavior and attitudes
- Use of time of all kinds
- Accuracy
- Safety of operations
- Scheduling of all kinds
- Customer surveys
- Sales behavior
- Customer satisfaction

Problems take many forms; four examples are given below. These illustrate items to be described in detail in a quality project report.

Bus Maintenance and Repair

- Time required to do a job
- Assignment of work
- Quality of repairs
- Parts available
- Schedule
- Buses to be repaired
- Mechanics available
- Training of mechanics
- What constitutes a quality job
- The checker of quality
- Repair, or buy new product?

Retail Trade Surveys of Customers

- Pleased customers

  What pleased?
  Reasons

- Complaining customers

  What complaints?
  Kind, frequency
  What can be done?

- Lost customers

  What didn't the customer like?
  Reason for shift
  What will bring customer back?

- Noncustomers

  What does the store lack?
  Reasons for buying elsewhere

Nursing Homes

- Are residents given proper food?
- Are residents assisted at mealtime?
- Is correct medicine given at proper time in correct dosages?
- Are nurses on call?
- Do residents receive proper medical care?
- Have there been incidents of stealing? What was done about it?
- Are residents' rooms safe?
- Do residents receive personal care: baths, hair care, clean clothes, clean beds, personal attention?
- Has abusive treatment ever been reported? What was done about it?
- Do residents have privacy?
- What caliber of help does the nursing home have?
- What happens to Social Security checks?
- What additional facilities and services does the home have: religious, books, magazines, radio, television, trips, birthday parties, tours, beauty shop?
- Is there fire protection?
- Are residents ignored?

A Laboratory

R. H. Noel and M. A. Brumbaugh of Bristol Laboratories reported the use of analysis of variance to discover sources of variation in a chemical assay, especially variation due to different analysts.[6] Four analysts were used, and their study led to the following conclusions:

> The experiment shows that analyst A is making some error in technique from day to day. Analysts B and C are doing poor work. Their cases require investigation. Analyst D is quite capable of doing this work, but appears to be careless.

Remedial steps were taken immediately. The authors concluded:

> There is scarcely any limit to the scope of the problems to which the statistical method is applicable. . . . Work has been done in the sampling of incoming materials, in sampling for the control department, in designing experiments for research and development (department), the use of sequential sampling in penicillin strain work, and in the routine control of processes and products.

6.   Train Workers for Quality Performance

There are many ways to train an employee, whether he is a worker or a manager. Examples follow:

- Set standards for new hires.
- Plan a schedule for semiannual or annual training.
- Train new hires.
- Train in any new and effective techniques.
- Provide special training for those with supervisory abilities.
- Train workers in methods of elementary statistics and how to use them on the job.
- Train in safety measures and how to avoid disaster.
- Train workers how to treat customers.
- Train in quality behavior and attitudes.
- Train in the prevention of errors.
- Train in better data collection and analysis.
- Train in saving all kinds of wasted time.
- Train to keep promises.

In transportation and health, the number-one quality characteristic is safety. The goal is zero errors. In local government, police, firemen, and first-aid personnel should answer calls within minutes. Concern for the customers (patients, accident victims, and fire victims) must come first.

Behavior and Attitudes

Service quality control is based on a face-to-face relationship between the customer and the salesperson. The behavior and attitude of the salesperson determines the quality of service, so special training is required of all employees. Examples of problems in this area follow:

- *Rudeness*—A salesperson sells a "package" and lets a customer know it; can sell only a "package."
- *Indifference*—A customer must wait until a salesperson has completed a personal telephone call.
- *Favoritism*—A salesperson deliberately waits on somebody out of turn.
- *Incompetence*—A salesperson doesn't know where the varieties of stock are.

To correct such poor services, teach employees to:

- Do the job right.
- Do the job without wasting time.
- Show interest in the job.
- Be polite and courteous to the customer.
- Be accurate.
- Be reliable—keep their promises.
- Treat customers fairly and justly—no favoritism.
- Understand the entire job so they can do a quality job.
- Do a quality job with lowest cost.
- Recognize differences between poor workmanship and quality workmanship.

Job Description and Learning

Job-description manuals—when they exist at all—can be too short or too wordy. Job descriptions should be in writing, and they should be discussed periodically with each employee to clarify points that are overlooked, misunderstood, or unclear.

In a job that requires a week or two to learn, random time sampling can measure progress toward zero errors by the use of improved methods. Caution should be used not to confuse a temporary plateau upon completion of learning with statistical control, providing an excuse that this level is "the best we can do." (See the section on the learning curve.)

7.   Institute Modern Methods of Supervision

A supervisor is responsible for maintaining quality or ensuring that quality is obtained higher up. The new supervisor must see that quality action is taken about such aspects of service as the following:

- Safe working methods
- Computer systems
- Processing methods
- Sampling designs
- Customer surveys
- Better training (e.g., Deming's and Juran's books and tapes)
- Clarifying terms in instructional manuals

- Distinguishing between defects of the system and special causes
- Recognizing faults in the system
- Stressing service quality characteristics
- Pressing for action at the top level
- Listening to employees

The supervisor reports in these areas should be acted on immediately. The supervisor can correct many system defects.

The supervisor can also act on many chronic problems affecting, for example, light, heat, air conditioning, computer problems (including placement of the equipment), word processing, placement of furniture, positioning of products, and design of data sheets, forms, and questionnaires.

Supervision—A Broader View

Supervision includes management, and there has to be a constant purpose in quality improvement. This means a new school of management in services, characterized by the following:

1. Cooperating at all levels—between and within departments
2. Sharing knowledge
3. Sharing decision-making
4. Encouraging the personal growth of every employee
5. Soliciting suggestions for improvements from every employee
6. Rewarding quality performance
7. Practicing participative management
8. Practicing management by leadership
9. Practicing management by trust
10. Emphasizing innovation
11. Training and retraining
12. Learning to listen
13. Stressing urgency
14. Listening to those who do the work
15. Giving the customer top priority

Modern management methods run contrary to the dictatorial methods still used in many companies today. Modern methods are based on trust and confidence, not on suspicion and fear.

Get rid of an inflexible organization. The top officials should mingle with the lower-level employees to discover problems and difficulties. They should cultivate a habit of listening. This means everybody will be involved in quality improvement. In a large organization, reduce the number of levels of organization, which are usually too numerous.

Quality performance should be rewarded, and top-level management should take an active part in giving out these rewards.

Top policy for services should be centered on the customers—their desires, needs, and preferences.

8.   Drive Out Fear

In a service organization there is an even greater and broader need to drive out fear. Fear is not an isolated event affecting only production workers in a factory; it is present in many customers and workers in a service organization.

- In health services (e.g., in hospitals, nursing homes), the patient fears:

  Surgery
  Activities related to surgery
  Theft
  Staff
  Nurses—RNs and LPNs
  Insane patients as roommates
  Dental work
  Cataracts
  Administrators
  Doctors
  Fire
  Abusive treatment
  Wrong medicine
  Wrong dosage*
  Neglect

- In transportation, fears of passengers and others include:

Note: One dose in seven is in error; the Health Care Financial Administration considers a 5-percent error rate acceptable. What is the death rate due to these errors?

Buses—breakdowns; accidents
Airplanes—ice on wings; pilot failure; faulty navigation
Trains—collisions, on the same track or through switching error; derailment
Construction sites—non-shoring; accidents; unemployment

These fears and others confront passengers, patients, and other workers—all customers of these kinds of services. In some situations these people need to be aware of what is going on. The patient wants to know the effect of a drug, the proper dosage, and the side effects, if any. The bus mechanic wants ample time to do a quality job. The customer wants to know sizes, colors, styles, and the store's services. The repair worker needs to know the stock and the tools, as well as where to find them, to meet the needs of the customers.

9. Break Down Barriers Between Departments

Teamwork within a service organization is necessary for a successful quality program. This will not occur simply by breaking down conflicts, quarrels, and differences of opinion *between* departments. There must be an acceptance of the need to avoid conflicts, quarrels, feuds, and jealousy *within* departments.

It may take years before the employees, supervisors, and managers can work as a team. The time it will take depends on the steps taken to overcome opposition.

There are several kinds of barriers to overcome:

- Between departments—between managers as well as supervisors
- Within departments

  Managers versus professional specialists
  Managers versus supervisors
  Supervisors versus employees
  Employees versus employees

You cannot break down barriers by instruction, because barriers are caused by many psychological factors, including:

- Jealousy
- Envy

- Ambition
- Fear
- Personality conflicts
- Differences of attitudes
- Fear of change
- Indifference
- Belief that a change will not improve matters

The barriers may be caused by different specialties of these managers. One approach to breaking down barriers is the following:

1. Interview each manager to find out in detail how each differs from the others.
2. Analyze the data for vital differences. Some of these may be merely differences in wording.

Here is how you might resolve differences related to emotional factors:

- *Jealousy and envy*—These people may "cooperate," even though their attitudes have not changed, because the boss believes in the program. Much jealousy or envy is kept well hidden. If it is not, one person will have to be transferred.
- *Ambition*—Opposition may arise from a misunderstanding. Point out that quality control will aid workers in their desire to advance.
- *Fear*—Much will depend on the nature of the fear. If the worker fears losing his job, point out that the job will get bigger and more interesting. The chances of improvement will be greatly enhanced.
- *The belief that "it won't improve things"*—By reducing errors and wasted time, productivity will be increased, costs will be reduced, employment will be assured, and everyone will gain.

Barriers are emotional and are made of strong stuff. These are attitudes developed over the years. You cannot change these emotions in a classroom the way you can change knowledge. We suggest several methods:

- Holding discussions showing how knowledge cuts costs without any capital investment
- Holding conferences with groups
- Holding conferences with individuals
- Setting up projects
- Using Deming's and Juran's books and videos

- Having Tom Peters talk about quality control
- Establishing a question box and taking it seriously
- Having a trusted person talk to the worker
- Emphasizing advantages: reduced costs, increased productivity
- Putting on a demonstration of improvement of quality
- Citing examples of improvement of quality in a field the worker is familiar with
- Using demonstration to apply continuous pressure for quality; a few years of this will help

One barrier is distinguishing between special causes and common causes. They are sometimes confused.

Special Causes

Special causes are faults that are caused by employees and that are to be corrected by them. Examples are:

- Typing errors
- Mathematical errors
- Errors employee can correct
- Faults in classification
- Misunderstanding rules, procedures
- Cash-register errors
- Faults in using credit cards
- Mixup of drug containers
- Misalphabetizing
- Providing wrong size, color, or item
- Incorrect billing
- Wrong sales slip
- Errors in making change (money)
- Computer programming errors
- Computer library errors
- Computer operator errors
- Mishandling shipments of books
- Faulty sampling instructions

Common Causes

Common causes are faults only top-level management is responsible for, or can correct. These occur in such areas as:

- Capital investment
- Budget

  Additional employees
  Training
  Travel
  Books and technical magazines

- Organizations
- Policies
- Reorganization
- Computer expansion
- Office furniture (bookcases, filing cabinets, etc.)
- Equipment
- Machinery
- Procedures, basic processes
- Sampling studies
- Customer surveys (sampling, nonsampling)
- Installing computer systems
- Credit-card systems
- Check-cashing procedures
- Customer conveniences
- Salary and wage scales
- Customer service

Common Causes That Can Be Corrected by Supervisors

Not all faults of the system have to be reported to top management; many can be corrected by the supervisors. These might include faults in:

- Heating
- Cooling
- Lighting
- Computers
- Word processing
- Typing
- Tabular forms
- Data sheets
- Mechanics of sampling
- Billing errors

- Claims-processing errors
- Filing systems
- Promotions
- Proposals from suggestion box
- Clerical errors
- Working conditions
- Negligence

System improvement can also be initiated by an employee: a new tabulation sheet, a new data form, a new calculation form, a better computer program, or a better sampling plan. Innovations and improvements are more likely to come from those who understand the work than from top-level officials.

10. Eliminate Numerical Goals, Posters, and Slogans

Slogans, targets, and exhortation have no effect on performance by service employees unless steps are taken to show how to achieve the goal. Two slogans that do not meet this criterion are IBM's

THINK

and another piece of art seen on an official's desk:

ACTION

They are just showpieces that caught the eye of a serious-minded manager.

Zero defects and zero errors are a different matter. In a service organization where one employee error can be dangerous, if not fatal, zero errors are a must. This is the case in medicine, transportation, chemical manufacturing, and construction work.

Examples have already been given or will be given to show that an employee error in health, transportation, nuclear power, or construction results in fatalities. No steps were taken, as far as the record goes, to prevent repetition of these errors.

Zero errors are required not only in potentially dangerous situations but in many others as well. Trends of errors should be driven to zero, not "standardized" at 5 percent, or 2 percent, or even 1 percent.

Where the errors of individual employees can have such devastating effects, the usual 85–15 division of responsibility for quality requires a drastic change. Responsibility of the employee varies from 50 percent to 100 percent, depending on the circumstances.

Setting a Target for Production

A private firm won a federal contract to process over 200,000 sample documents. The clerical staff was a mixture of housewives and high school and college students. With one exception, this was their first job that required a training course using a manual of over 100 pages. This was not a simple clerical job.

The vice president set a production goal of 2000 documents processed daily. This took the form of a slogan, which he held to firmly, even though the learning curve showed production of less than 500 per day at the start. The 2000-document daily production goal was not reached until the 18th day with 14 people. This was a closely supervised operation with an excellent manager in charge.

A working lot of 300 documents was subject to a sample inspection of 50. A rejected lot was returned to the person who produced it. Then 100% of the lot was examined. This method kept the error rate at a low level.

The setting of a goal did not speed up production, because the job was too difficult. For all intents and purposes, the prescribed goal was ignored as being unrealistic and unreachable. It simply confused the workers and the manager.

The slogan "Friendly service is our style" is prominently posted in a supermarket. "Friendly service" comes from the people waiting upon you; either they already have it or they do not. No sign, no matter how prominently placed, is going to change the unfriendly person.

Friendly service has to be demonstrated in a classroom, on the job, or in a particular situation. It cannot be taught by a slogan, however clever the slogan may be.

Friendly service in a supermarket could take the form of remarks like the following:

1. "Hello. How are you today?"
2. "Do you want these in a paper or plastic bag?"
3. "Have a good day" or "Take care."

The phrase "is our style" seems to mean "is our practice" or "is practiced all the time," or its equivalent. The word "style" is ambiguous; "habit" would be clearer.

11. Eliminate Work Standards That Prescribe Numerical Quotas

Workers are subject to work standards and standard times, which were developed by industrial engineers. Time and motion studies, plus allowance for personal time, plus other allowances, give "standard time."

There is no need for "standard time."* The use of modern random time sampling (RTS) will describe the actual situation: working time as well as idle time, time waiting for work, machinery down time, and other wasted time. RTS will give the following:

Working time	Frequency	Cost
Idle minutes	Frequency	Cost
Time waiting for work	Frequency	Cost
Down time	Frequency	Cost
All others	Frequency	Cost

In this way, cost of non-quality can easily be measured, and steps can be taken to reduce if not prevent it. The losses are eliminated.

This assumes RTS is used as described, including the minute model. This gives measurements as well as frequencies. The tour and ratio-delay methods are obsolete.[7] They lead to unsound sampling and yield only questionable frequencies of occurrences.

Table 2.2 summarizes an actual 10-day random time sample.

This information is based on a probability sampling covering 8 hours every day for 10 days for a total of 80 hours. The sampling rate is four samples per day. The last two columns give the estimates for the population for hours and labor cost.

If "waiting for work" cost $50 for 10 days, or $1200 a year, it would pay to investigate whether the causes were random or a departure from random. Idle time and all forms of wasted time can also be measured, as

---

*The author came to this view in 1955.

Table 2.2
Summary of 10-Day Random Time Sample

Work Status	Sample Employee Minutes	Percent	Estimates	
			Employee Hours	Dollars
At work	1227	83.4	2454	7683
Other	108	7.3	216	676
Waiting for work	8	0.5	16	50
On leave	129	8.8	258	808
Total	1472	100.0	2944	9217

can departure from safe rules of operation. These and any other departures from quality performance can be detected and prevented.

RTS has been used to solve the following cost problems:

- Loading of freight cars
- Unloading of freight cars
- Allocation of different activities to different funds
- Division of labor of one operator working on two machines simultaneously
- Cost of time to do "extra jobs" not assigned to the division
- Waiting for work
- Division of time for different groups to perform several activities

Tippett's snap reading method, Morrow's ratio-delay method, and Waddell's work sampling using the tour method have all been superseded by random time sampling derived from the theory of probability sampling and the use of the minute model.

The advantages of RTS are many; it can be used to:

- Estimate percentage of lost time and idle time
- Determine money cost of lost and idle time using sample method of estimation
- Measure decomposing time and cost of joint activities
- Cost specific worker and machine activities

The major difficulty is not sample design, but obtaining objective and operational definitions of the activities to be observed so that identification is highly reliable. The method of estimation is that built into the sample design; there is no need for any arbitrary method of estimation.

This method is described in Chapter 3 in the section on random time sampling. The principle, sample selection, and tabulation and analysis of the data are described in detail in the author's *Applications of Quality Control in the Service Industries* (Chapter 24)[8] and *The Quest for Quality in Services* (Chapter 15).[9]

Piece Work—An Example of Numerical Quotas

In a factory, some of the small punch-press work was paid at a piece rate. The workers on piece rate gauged their production by watching a counter. When it showed a certain number, they shut down the machines and took a break. Productivity suffered. They could have earned more than the boss but didn't dare to. If they did, the piece rate would be reduced. This shows the futility of numerical quotas.

12.   Eliminate Barriers to Pride in Workmanship

In a factory, workers can feel pride in the product or part of a product they produce. However, there is no physical product in a service organization. In the following service industries, pride may mean doing something that pleases the customer.

- Banks
- Retail stores
- Insurance companies
- Government offices
- Providers of personal services (laundry, beauty shops, barbershops, cleaning services, hotels, and motels)
- Businesses
- Transportation
- Public utilities

In health care, pride is helping the patient get well. In a nursing home, pride is treating patients with care and compassion, as civilized human beings.

In any service organization, pride takes the following forms:

- Pleasing customers
- Doing things on time
- Keeping promises
- Calling by telephone if promised
- Going out of your way to satisfy the customer
- Solving a customer's problem
- Doing an excellent job for the customer

These can't be taught in the classroom. One's personal conduct is a factor. Face-to-face contact with the customer may increase the employee's responsiveness.

13.   Institute a Vigorous Program of Education and Retraining

Quality control requires a continuous program of both statistical and nonstatistical information.

Many employees of service organizations do not understand simple mathematics, cannot read or follow instructions, and will not stay with the job (leading to turnover).

Curriculum

The training courses available range all the way from decimals to calculus. In what order will these courses be presented? Who will make the decision? Where will the content come from? How will the courses be related to the work of the employees? When does someone need retraining? How will "massive" education be defined? How will mastery of the knowledge be tested? Who will see that "massive" education is planned, directed, and conducted with any degree of effectiveness? How difficult will the courses be? Setting up a series of courses will not be easy.

The Teacher

Neither manufacturing nor service organizations hire anyone who is specifically a teacher, let alone a good teacher. Where will the teachers of company programs come from? Some possible ways are:

- Find out whether any employee is a former teacher.
- Hire a consultant who can teach.
- Train an employee who can teach.
- Hire a statistician who knows the subject matter (an M.A. degree is sufficient) and knows how to teach.
- Let anyone who is interested teach. However, this may raise a serious question about the value of such a course.

Teachers not only have to teach; they also have to test the success of their teaching. The process of teaching is not as easy as it seems. Deming says it is hard to get a good statistician. It is also hard to get a good teacher anywhere.

These reasons show why a "massive" education is difficult, why it has obstacles to overcome, and why it is easy to talk about but very hard to implement.

The Employee

The ability to learn varies widely among employees. In service organizations one has to start with the elementary schools, where courses and topics include:

1. Decimals, fraction, percentages
2. Measurement of area and volume
3. Algebra
4. Elementary statistics
5. Geometry
6. Analytical geometry

No employee is expected to go through all six courses. Two elementary courses plus elementary statistics with the emphasis on quality control will be enough.

Communication

This should be part of training.

Messages Transmitted by the Top Officials

- Policy statement
- Recognition days announced

- Quality improvement plans for following year
- Reorganization changes
- Company-wide announcements
- Training courses

Messages Sent from the Workers

- Suggested improvements
- Complaints about conditions of work
- Job descriptions and assignments
- Clarifying instructions
- Training courses

14. Put Everyone to Work Bringing About this Transformation

Make quality control everybody's business every day. Extensive education will help bring this about. This means a training session is given to each new employee. It means training and retraining regular employees, with special attention to safety.

Remember, customers don't buy any services from the CEO. They don't buy anything from a manager. They buy from salespersons; they do business with clerks; they make inquiries of other clerks. These people determine the quality of service the customer receives.

It will take at least five years to reach maturity—to the point where everyone believes in, and works on, quality improvement. One answer is a quality budget that includes:

1. Departmental quality budgets, projects, and costs savings
2. List of long-term improvement projects
3. Reduction of turnover: rewarding producers and innovators; rewarding quality producers
4. Quality orientation of all new employees
5. Quality projects for the coming year

Create a management structure that practices quality improvements daily. This requires statistical capability of a special kind, which Deming describes in his latest two books. They illustrate the relatively simple statistics needed. The major problem is to learn how to use these statistics effectively.

One must learn business operations as they affect the job and general knowledge of the business if he or she is going to contribute to quality improvement.

A statistician's duty is to discover, study, and analyze statistical problems that lead to quality improvement. The job of the statistician is to identify management's problems that are statistical or have statistical aspects and then help solve them. This will aid in the transformation.

Table 2.3 condenses Deming's 14 points.

Table 2.3
A Condensation of Deming's 14 Points

Point Numbers	Content
1 and 2	Constant purpose for a new age
3 and 4	Mass inspection, price tag
5	Find problems
6, 7, and 13	Education and re-education
8, 9, and 12	Fear, barriers, pride (psychological aspects)
10 and 11	Numerical goals, work standards
14	Continuity of program: everyone working on quality improvement

These groupings combine related topics without change. Items 8, 9, and 12 are psychological and not at all similar to "quality control statistics." They require a different approach.

CHAPTER 3

A Quality Program for Services

This chapter outlines a quality program that starts with the customer and the employees, then follows the problem survey and the training needed, including the development of teamwork. Data collection with quality as the goal is developed, with statistics being given its proper place.

Then non-quality characteristics and their prevention are discussed, including errors, delays, wasted time, safety, and defects in purchased products.

The importance of a suggestion system and the learning curve are pointed out as important parts of a quality plan. Finally, there is a section on the cost of non-quality characteristics.

1. Survey Customers' Requirements

A company succeeds by meeting the requirements of customers. A survey is used to determine these requirements. A survey takes the following forms: (1) 100-percent coverage and (2) several fractional samples that include 100-percent coverage every year. This kind of survey is for the customer-oriented organization.

Small Surveys

If the total number of customers is 500 or less, 100-percent coverage is possible.

1. Include all customers. Use the mail or personal interviews.
2. Divide them into categories: pleased, contented, indifferent, dissatisfied, or lost.
3. Repeat study every three months.
4. Find out what pleases, why some customers are indifferent (one may lose these), why they are dissatisfied (one may lose these), why they are content (one can rely on these), and why they have been lost.
5. Take appropriate action. Quality improvement is needed.

Large Surveys

If the number of customers is large, random sampling is required; if data are available for grouping, stratified sampling is required.

1. Rotate sample so all customers are included every six months.
2. Divide them into categories: pleased, contented, indifferent, dissatisfied or lost.
3. Determine what quality services they like and dislike.

   - What pleases
   - Why they are content
   - Why they are indifferent
   - Why they are dissatisfied
   - Why they have been lost

Take action on the indifferent, dissatisfied, and lost customers. Improve quality of services.

Customer surveys are a continuous program providing the company with basic information. Customer complaints may be only the tip of the iceberg.

Customers are the key to quality. Customer surveys allow the company to keep in constant touch with customers' needs, requirements, preferences, attitudes, and ideas. Develop a quality-improvement program based on these characteristics.

Discovering and Eliminating Customer Complaints

Department stores prevent stealing by using a device attached to clothing that rings a bell to identify a shoplifter. Sometimes, when customers make a purchase, the device is not removed and the bell is ignored. Customers then have to return the clothing to get the device removed, since a special tool is required.

This happened to a customer living in Wichita when she made a purchase in Kansas City. She had to return the item to get the safety device removed. She had a sales slip showing that she bought it; otherwise she might have been accused of theft. All customers face this same potential situation.

Another complaint is inaccuracy. A department store marked down the price of a garment for quick sale. The price was not legible, so the salesperson charged a customer more than the original price. The customer discovered this and showed the garment to the clerk, who found that the customer had been overcharged $27.

Other complaints are that a salesperson ignored the customer; lack of salespersons to check out a purchase; and that the salesperson sent the customer to the customer service department.

The goal is to put an end to these errors and inconveniences through periodic customer surveys; otherwise the company will lose customers.

Example

The Internal Revenue Service conducted a nationwide sample for audit purposes. It was the first time this had been done. The estimate from a sample of 160,000 returns was $1.5 billion in error. Business returns had twice the error of non-business returns.

This information provided a basis for the new few years' operations. It showed where most of the error was by locality, type of business, and other factors.

This study revealed for the first time a quantitative measure of non-compliance. It showed that taking an alien as a personal exemption was an illegal practice. Many cases of this type were found in large cities, such as Boston and New York.

This study was an enumeration used strictly for analytical purposes. It was assumed that a taxpayer's habits did not change from one year to the next.

As a result of the feedback from this sample, a new schedule on personal exemptions was added to Form 1040 (it is still there), and the instructions were made more detailed.

Example

The Interstate Commerce Commission (ICC) conducts nationwide sample studies to estimate circuity factors and to obtain key information on household movements by interstate truck lines.

Circuity factors are used in cost work. The circuity factor is actual miles divided by short-line miles, the distance requiring no transfer of lading. Since actual miles equal or exceed short-line miles, a circuity factor is the percentage by which the former exceeds the latter. These factors change very slowly, so the data are used for several years.

The interstate household goods study was conducted to reveal causes of customers' complaints. Estimated cost, time of delivery, and insurance claims were the principal complaints. This study gave the ICC a factual basis for improving moving practices: cost variations, delays in movements, settlement of insurance claims, and extent of damage.

Customers Buy a Sample of One

Customers buy a sample of one, whether the purchase is in a retail store, supermarket, bank, insurance company, telephone company, airline, gas company, electric utility, or automobile dealership.

Quality is unknown; a customer buys in the dark. Estimation of quality of a purchase is based on experience, a salesman's talk, advertising, or other persons. There is no mean, no variance. In all such cases:

- There are no quality control records.
- There are no written records.
- Quality data are unknown.
- Warranties do not measure quality.

Quality of a utility is accepted; you hope it will be reliable.

Quality of products and services, including maintenance, repairs, efficiency, and length of life, has to be appraised by the customer over time.

Customer Perceptions of Quality Based on Limited Information

What does "customer perceptions" mean?

What does "limited information" mean?

Does "customer" mean the ultimate customer? The householder?

Does "perceptions" mean "percepts"?

Does "limited information" imply information is withheld?

We greatly underestimate the ability of the customer to buy quality products in retail stores. We also miscalculate the ability to buy quality services.

The customer's perception of quality of a product is different from that of a manufacturer.

The customer is *not* concerned with the quality of thousands of parts of an automobile. The customer wants

- Reliability
- Low operating costs
- Low repair costs
- Low maintenance costs
- Dependability
- Satisfaction
- Minimum of troubles
- Product or service that meets requirements

Customers learn from experience. No sales talk, cashback, refund, or other gimmick will sell a product the customer doesn't want. Customers are *not interested in production*. They want to know how the product *operates*. They want to know how the product works. They want a product to operate reliably with a minimal amount of maintenance and repair.

They want from a repair shop or dealer the following:

- Courtesy
- Accuracy
- Reliability
- Efficiency
- Reasonable price
- Timely action

Questions to Ask

Pleased Customers

1. What pleased you most?
2. Which products did you like?
3. Which services did you like?
4. Which products need improvement?
5. Which services need improvement?

Dissatisfied or Complaining Customers

1. What didn't you like?
2. What products displeased you?
3. What services displeased you?
4. What needs to be changed?
5. Have your complaints been corrected?

Contented Customers

1. Are you content with our products?
2. Are you content with our services?
3. What improvements, if any, are needed in our products?
4. What improvements are needed in our services?

Indifferent Customers

1. Are you satisfied with our products?
2. Are you satisfied with our services?

Lost Customers

1. What didn't you like about our products?
2. What didn't you like about our services?
3. What improvements would you suggest?
4. Do you consider our prices too high?
5. Do you consider our quality too low?
6. What changes would bring you back?

Noncustomers

1. What are your reasons for buying elsewhere?
2. What changes need to be made for you to become a customer?

Higher quality is needed, so an improved quality program is required.

Companies That Are Not Customer-Driven

There are many more companies that are profit-driven than are customer-driven. The following describe companies that are not customer-oriented.

1. They sell only that which is profitable. This means stressing the needs of those under 40 years of age and neglecting the elderly. It means ignoring smaller- and larger-than-average sizes in clothing and shoes. It means catering to young people who live on french fries, hamburgers, cheeseburgers, and pizzas.

2. They discontinue carrying products some customers like.

- Saffola margarine—replaced by Weight Watchers margarine
- Sodiphene antiseptic
- Unguentine salve for burns
- Tincture of iodine—disinfectant
- Leather shoes for women—foreign-made
- Women's petite clothing
- Clothes to match the season's weather—no summer clothes in January, nor furs in August

3. They do not carry products or provide services needed.

- Do not carry preferred brand
- Do not carry correct sizes
- Do not carry desired composition of a product
- Have no maternity department in department stores (in one case, in a city with a population of 60,000)
- Do not design hairstyles specifically for the elderly

4. They sell automobiles by appealing to interest in price and speed, not quality—offering, for example:

- $2000 cash back
- 0% or 2.9% interest on loans
- $1000 bonus

The customer-driven organization does extra favors for the customers:

- Writes thank-you notes to customers who make purchases
- Orders nonstandard sizes of clothes and shoes (sizes that fit)
- Caters to the elderly
- Accepts returns of goods that are not satisfactory, and replaces them with items that are

2.   Conduct an Employee Survey

An attitude survey should be companywide, to determine where everyone stands on quality. Many employees will not have heard of the word; many believe they are already doing a quality job; and very few will know the true meaning of the word.

This requires several meetings aimed at explaining the significance of the term:

- Prevention of errors
- Reduction of waiting time
- Reduction of delay
- Reduction of idle time
- Courtesy, kindness
- Polite handling of complaints

all of which will reduce costs.

The questions asked in an employee survey may include:

1. What does "quality" mean to you?
2. Give examples of "quality."
3. What plans do you have to advance "quality"?
4. Have you any examples of "quality" improvement?
5. Have you and your fellow workers ever discussed "quality"?
6. Has your supervisor ever talked "quality" to you, or to his/her staff?

Very few will understand "quality." The first group to convince is top managers, followed by middle managers, supervisors, and employees. Full cooperation is required for a successful quality-improvement program. Teamwork is essential.

Managers and employees have to be convinced of the importance of a quality-improvement program. Many of the suggestions for improvement will come from the employees who are doing the work. This is why proposals for improvement are canvassed companywide, and continuously. Unless key managers accept quality as a vital and continuous process, it will not succeed.

What kinds of plans will result from this survey?

1. Form a quality council.
2. Form quality groups headed by a supervisor or worker.
3. Develop quality teamwork.
4. Start training classes emphasizing self-respect, pride, and trust.
5. Encourage employee cooperation and teamwork; employees take a detailed ownership role in operations.

Quality groups may be formed as quality circles. Workers form a working group led by a supervisor. They select an important problem to work on. They apply one or more of six simple statistical techniques, in addition to a cause-and-effect diagram that is not statistical. The statistical techniques are as follows:

1. A *histogram* is a frequency distribution.
2. A *checklist* is a tally sheet.
3. *Pareto analysis* is arranging factors in order of percentage.
4. A *scatter diagram* is a relationship plot.
5. *Stratification* means the same as Shewhart's "rational subgroups."
6. *Control charts* are needed to detect statistical control.

Members of these circles will most likely not understand simple statistics—hence the need for continual training. Select a promising group to make a *pilot study* to test the method:

1. Choose a group of 10 or fewer.
2. Teach them statistics first.
3. Examine several problems.
4. Select a problem.
5. Work on it together.
6. Set standards.
7. Measure results.
8. Make a final decision based on savings.
9. Select another problem.

Stress teamwork and 100-percent participation.

Behavior and Attitudes

In most services, quality is performance, and is determined by the person behind the counter or desk.

For quality treatment, the customer can expect:

- Attention as soon as possible
- Courteous and polite treatment
- Undivided attention
- Careful consideration of what the customer wants
- Attention to complaints
- Resolution of complaints to the customer's satisfaction
- An expression of thanks (whether the customer buys anything or not)
- An accurately filled-out (or read-out) sales slip
- Responsiveness to a question
- Explanation of procedure
- Acceptance of full responsibility for a personal or company error

Avoid excuses. This approach follows the practice of Nordstrom's department stores. They accept any item as a return, whether they carry it or not. Nordstrom's put customers first.

How do you handle fear, barriers, and pride? We have already pointed out that special teaching methods must be used, not those in a conventional course involving knowledge. The most receptive employee will be a different kind of person from the common type of service personnel. This person should combine a certain set of attitudes, an understanding of the job, and a desire to grow in quality improvement.

Such a person must *not* have any of the following traits:

- Rudeness
- Aggressiveness
- Discourtesy
- Negativism
- Inaccuracy
- Untrustworthiness
- Unreliability
- Anti-customer attitudes
- Uncooperativeness
- Cynicism
- Carelessness

The following remark lost a customer: "What are you getting so hot about?" (an answer to a question about the cost of gasoline service).

Employee training is needed just as soon as the employee is hired to avoid these anti-quality attitudes. Do not hire those with anti-quality attitudes.

Ignoring the customer is the worst non-quality characteristic of them all.

Barriers

Cooperation *within* and *between* departments is necessary. There is a need to break down barriers between workers, technical professionals, supervisors, and managers. This is *within* a department. There is also a need to break down barriers *between* departments: market research, sample design, training, design of service components, production, automatic surveys, customer service, and the computer system.

Breaking Down Conflicts

Fear, barriers, and pride in workmanship cannot be broken down in any ordinary classroom or seminar. They are *emotional,* not just intellectual.

Emotions cannot be changed the way knowledge can. Suggested methods for bringing about change are:

1. Get a trusted person to discuss the change.
2. Use a group of people who approve of the change.
3. Discuss the advantages of this change.
4. When some of the task force accept change, others begin to accept it half-heartedly; take advantage of this.
5. Acceptance may occur because top management started the change.
6. Change supervisors or managers.

3. Conduct a Problem Survey

A problem survey is the heart of quality control: finding problems to solve. Examples are:

- Look for errors to prevent.
- Look for ineligibles who are paid and eligibles who are not paid.

- Look for storage that leads to errors.
- Save paper and people by sampling.
- Work on a problem that saves *time*.
- Work on a problem where *safety* is ignored.
- Work on a problem where *probability sampling* cuts costs.
- Work on improvement of customer service.
- Survey lost customers to discover why they left and how to get them back.
- Study how to improve the learning curve in training and on the job.
- Examine paper forms for improvements.
- Improve suppliers' quality control.
- Study how to prevent customers' complaints.
- Improve company's shipping system.
- Inspect automotive fleet; have quality improvement done in the factory.
- Improve training courses.
- Take an inventory of computer projects to find where improvements can be made.

A longer list follows.
How will this survey be made? By:

- Talking to managers
- Studying returns from the suggestion box
- Administering a short questionnaire
- Presenting a series of conferences
- Talking to employees
- Using successive samples if the organization is large
- Surveying publications
- Discussing experiences

Problems Revealed by the Survey

Some examples of the kinds of problems uncovered in such surveys are given below (authors in parentheses).

1. Lack of training, lack of instructions, deviation from procedures

   - Did not know how to run test equipment (Grant)

- Inspector errors (Grant)
- New inspector unfamiliar with work (Grant)
- Errors in laboratory measurements (Grant)
- Operator deviated from procedure (Grant)
- Chemical analysis differed widely on same test (Brumbaugh)
- Difference due to operators, not to machines (Grant)
- High error rate among coding clerks in data processing (Rosander)
- High error rate among keypunch operators (Rosander)
- High error rate on original documents (Rosander)
- Computer programmer used wrong estimating equation (Rosander)
- Computer programmer used wrong equation for standard error (Rosander)
- Computer programmer included class of persons to receive federal benefits that should have been excluded (Rosander)
- Computer operator neglected to run four reels of tape in a tabulation (Rosander)
- Labels were switched on bottles of two drugs (FDA)
- Wrong label; misbranded (FDA)
- Wrong count of pills in bottles (FDA)
- Transposition label error: 8.3 milligrams instead of 3.8 milligrams (FDA)
- Deliberate falsification of data in two cases (Rosander; federal report; press)

2. Tests unsound, inadequate, or questionable

- Error in plan for testing two shipping containers (Purcell)
- Laboratories disagree significantly on tests of same sample (Youden)
- Buyer's and seller's tests of same product differ (Grant)
- Unsound experiment for testing plutonium at two different sites (Rosander)
- Massive dosage testing of chemical additives such as cyclamates questioned as sound technique (scientists writing in *Science*)
- Simulated tests of automobiles for miles per gallon of gasoline (several sources, including EPA)
- Textile-mill manager uses test data from one operator; sample of one (Rosander)

3. Sampling procedures wrong or inefficient

- Military Standard 105 used for variables in receiving department (Rosander)
- Lower incoming fraction defective needed for sampling to be effective (Rosander)
- Sample very inadequate to estimate pilferage in a large chain of retail stores for federal tax-adjustment purposes (Rosander)
- Inefficient acceptance sampling plan (Rosander)
- Judgment sample used instead of random sampling in acceptance sampling (Rosander)
- Tour sampling used instead of random time sampling in work analysis (Rosander)
- Judgment sampling used instead of random probability sampling in testing and in making estimates from carload and shipload receipts (Rosander)
- Use of 100-percent sampling instead of a properly designed sample (Rosander)
- Excessive automated measurements on characteristic of air quality: 300 overlapping measurements on every 500 cubic centimeters of air (Rosander)
- Use of judgment instead of random probability sample in market survey (Rosander)
- Use of random instead of time-ordered sampling in process control (*Quality Progress*)
- Need to redesign sampling for control of high-speed production (*Industrial Quality Control*)

4. Process changes deliberately introduced or unknowingly changed

- Change in heat treatment procedure and method (Grant)
- Change in pH level (Grant)
- Operator adjusted machine (Grant)
- Change in vendor or vendors (several)
- Change in shifts (FDA)

5. Machine trouble (several sources)

- Tolerances too narrow
- Tolerances too broad

- Machine capability studies needed
- Tool wear
- Machine needs adjustment
- Machine needs repair

6. Materials and product trouble

- Material not strong enough (Rosander)
- Material cannot meet specifications (several)
- Poor design (several)
- Defective assembly (several)
- Bad can seams (FDA)
- Defective pacemakers because of design, assembly (FDA)
- Unacceptable manufacturing practice (FDA)
- Too many rejects (several)

7. Inspection trouble (Deming's point 3)

- Gauges in error (Grant)
- Inspection errors (Grant)
- Practice of "pulling in" at specification limits (Juran)
- Inspector fatigue (Grant)
- Variation among inspectors (Grant)
- Quality-assurance inspections omitted (FDA)
- Process of inspection for certain defects adds more defects (*Industrial Quality Control*)
- Faulty sample plans and selection of samples (several)

8. Control problems

- Wrong characteristics controlled (Grant)
- Separate control charts needed (Grant)
- Excessive weight or fill above or below specifications (ASQC Transactions, *Industrial Quality Control*)
- Too high fraction defective from vendor (ASQC Middle Atlantic Conference)
- Uncontrolled situation looks controlled because of sample bias (*Industrial Quality Control*)
- Lack of control over quality of input data (Rosander)
- Lack of control over computer operations (Rosander)

- Sample of four or five for $\bar{X}$, R charts selected from different populations by using one from each filler, needle, turret, pad, etc. (Grant, *Industrial Quality Control*)
- Ignoring time order of sample values for $\bar{X}$, R charts and other process-control charts (*Quality Progress;* other)

9. Safety problems

  - Sharp edges not smoothed
  - Materials not strong enough
  - Electric wiring subject to rubbing against metal
  - No guards covering moving parts
  - Inadequate protection against hazards: fire, explosions, shock, acids, gas
  - Lack of employee safety devices: goggles, safety belts, gloves, respirators, shoring, scaffolding
  - Wheel fastened to axle only with nut, so rotation runs wheel off axle

10. Random time sampling problems

  - Random time sampling (work sampling) applied to inspection
  - Random time sampling (work sampling) applied to utilization of machines to measure, cost out, and reduce down time
  - Random time sampling (work sampling) applied to utilization of equipment
  - Random time sampling (work sampling) needed to cost out time spent on scrap and rework

11. Computer problems

  - Failing to make proper feasibility study to buy a cost-effective computer and design and implement an efficient computer system
  - Putting large volume of data of limited use, or of little or no use, in computer memory (McClure, "Empty the Computer," *Quality Progress,* September 1975)
  - Programming computer for unnecessary statistical analysis of $\bar{X}$ chart data (*Quality Progress,* September 1977)
  - Using computer in a way that removes quality control from

factory floor or production line and introduces delays in corrective action (Rosander)

- Processing data and data analysis in ways that may take hours and delay taking action on $\bar{X}$ chart showing out-of-control and increased quality costs
- Failure to apply quality controls to computer operations to input data, entering input data into computer, programming, running tapes or disks, or to printouts (Rosander)
- Confusing sample audit and computer calculation of average quality level with quality control (*Quality Progress*, Rosander)

12. Analysis of quality control data

- Neglecting to use analysis of variance to test for common mean which is evidence of control (Rosander)
- Neglecting to use analysis of variance to test variation between and within inspectors and others to determine amount of variation and to reduce or eliminate it (Rosander)
- Neglecting to use analysis of variance to test variation within and between laboratory tests (Youden)
- Neglecting to use analysis of variance to test variation due to rational subgroups to determine whether more than one control chart is necessary (Rosander)
- Neglecting to use analysis of variance to determine if an apparently in-control situation is really out of control due to bias (*Industrial Quality Control*)
- Neglecting to discover out-of-control situation within control limits due to runs, trends, cycles, or sample bias (*Industrial Quality Control*)

13. High-speed production, including automation

- Failure to adapt quality-control techniques to high-speed production (*Industrial Quality Control*)
- Failure to design an adequate quality-control system for high-speed production or automation
- Failure to adapt a computer to high-speed quality control
- Failure to use computer or special equipment to design and implement high-speed quality control

- Failure to understand how a system should be designed and implemented in high-speed quality control
- Failure to design a high-speed production-control system so product conforms to specifications

4. Start a Program

One way to start a quality program is to select a problem or project where the outcome will be successful. Choose:

1. A simple problem that needs solution
2. A unit or organization that accepts quality improvement
3. A unit or organization that is willing to master the relevant statistical techniques
4. A project whose money savings can be calculated and are substantial

This could involve any unit or department in an organization—a group under a supervisor. The most important traits of this group are teamwork, cooperation, and the ability to work together on quality improvement.

Possible projects include:

- Number of errors
- Error rates
- Defects in purchased products
- Delays
- Other lost time or excess time
- Failures
- Customer complaints; need to restore discontinued merchandise

There may be no need to change the organization chart, since employees are the critical factor. Their attitudes, training, experience, motivation, and cooperation are the key elements.

A second way to start a quality program is to organize quality circles. At the start there may be only a few of these circles. The supervisor is usually the leader. Members are supposed to learn how to apply seven techniques:

1. Tally
2. Histogram
3. Pareto analysis
4. Stratification
5. Control charts
6. Scatter diagram
7. Cause-and-effect diagram

Joining a quality circle is voluntary at the start, but the goal is to make membership compulsory.

Quality circles require careful planning and preparation. A weakness is that they are restricted to local, specific problems. If they meet a chronic problem, only top-level management can solve it. The method often calls for a facilitator who works in different circles to act as a technical instructor.

Quality circles won't succeed without management support, teamwork, and statistical capabilities.

A third method of starting a quality program is to form a quality council at the higher levels of management. This may cover the entire organization or only a part of it. The quality council assumes full responsibility for quality improvement. The steps are:

1. Quality planning
2. Quality control
3. Quality improvement

All of these approaches have as a foundation the results from the:

1. Customer surveys
2. Employee surveys
3. Problem surveys

Quality planning requires setting quality goals, listing and selecting problems, measuring progress, and taking action for quality improvement.

Statistical control means control charts and correcting specific defects: np, p, c, u, $\bar{X}$, R charts, and the line graph.

Quality improvement calls for elimination of specific causes, which is the responsibility of the worker, and elimination of chronic defects, which only management can do.

A modification needs to be made in Dr. Juran's trilogy.[10] Quality control, instead of maintaining the status quo, may result in quality improvement.

1.   Wade Weaver of Republic Steel reports that the company's use of quality-control charts in the steel industry, and the corrective steps that arose as a consequence, resulted in a savings of over $175,000 in one year.[11] This example involved the manufacture of steel ingots that were designed to weigh 5300 pounds but that varied from about 5500 pounds to about 5800 pounds, by lots. After using statistical-control charts for about one year, the process was brought under better control, so that at the end of this time the ingots averaged 5296 pounds, with an average variation from one ingot to another of only 82 pounds. This reduction in variability meant very substantial savings for the producer as well as a more uniform product for the consumer. This was a big cost-saving breakthrough.

2.   At a quality conference held at the University of Maryland, the representative of a vegetable-canning establishment described how they were canning sauerkraut. They finally introduced statistical quality control and discovered they were filling, on the average, 14–16 ounces per can.

The can was labeled "12 ounces" so they were giving away 2 ounces on the average with every can. The amount of this loss was never calculated, but it was substantial.

The method of packing was changed so the average was reduced to 12.5 ounces, with no can weighing less than 12 ounces.

3.   There are many examples of error rates and defective rates being driven toward zero. Examples were cited earlier by Wiman.[2] See *The Quest for Quality in Services* for additional examples.[9]

5.   Provide Total Training

Total training includes everyone from the CEO to the lowest-level worker. This training should be graded.

• Top-level management

How to manage for quality
How to plan for quality improvement
How to implement for continuous quality improvement
How to establish customer policy

- Middle management

  How to manage for quality
  New technical methods
  Changes in supervisory methods
  Teamwork and cooperation
  Working with line people
  Improved quality data
  Implementing customer policy

- Supervisors

  New technical methods
  New approaches to personnel
  Methods to obtain quality data
  Rewarding employees
  Zero defects
  Attitudes and behavior toward customers

- Employees

  New technical methods
  Refresher course on methods
  How to work toward zero defects
  Teamwork and cooperation
  Attitudes and behavior toward customers
  Improve data collection and processing

Hiring Policies

The policy to follow continuously is to hire the best applicants, that is, applicants who:

- Can read, write, and understand instructions, both written and oral
- Understand elementary arithmetical concepts such as the average, range, percentage, fractions, and decimals
- Are promotable
- Show managerial potential
- Have mastered statistical quality-control techniques
- Know how to talk properly over the telephone and who do not make decisions they are not supposed to make

Employee Ideas

Employees who do the work see defects, errors, failures, wasted time, and other characteristics that need to be changed, but they usually stay silent. Employees can be the source of many improvements if encouraged to present their ideas for change. A few companies that mine the ideas of employees have found that it pays off in reduced costs and quality improvement. Top-level management needs to encourage employees to make suggestions for continuous quality improvement.

Public Education

It is estimated that business and industry are spending $25 billion a year to educate workers in subjects they should have learned in elementary school. There is a crisis in education because schools are not preparing students for: 1) being informed citizens, 2) having a career, and 3) attending college. The educational process consists of a teacher imparting knowledge to a class of students; everything else is supportive.

A core curriculum will include concepts of elementary arithmetic: long division, fractions, decimals, percentages, averages, range, areas, and volume. These concepts and others are not hard to teach to a class if the teacher understands what the concepts really mean.

This kind of arithmetic ability is needed to meet the quality-control requirements in a factory and to match foreign competition. This is a necessary, but not a sufficient, aspect of elementary education.

6.   Train Salespersons, Clerks, and Attendants

For quality performance, salespersons, clerks, and attendants should have the following traits:

- Courtesy
- Accuracy
- Reliability
- Ability to listen to a customer

Courteous behavior consists of polite, helpful, or considerate acts. The employees are polite and helpful to the customer. They find what the customer wants. They know the stock, files, records, and documents. They answer the customer's questions in a thorough and accurate manner. They listen to the customer's complaints and solve them, or take him to a person who can solve them.

They are friendly. The customer is greeted with "How are you today?" or something similar. On leaving, the customer hears "Have a good day" or the equivalent. The whole atmosphere of the business recognizes the customer as an important person.

Accuracy is necessary in employees' mathematics, spelling, and grammar. All arithmetic should be correct, because methods of verification are easy for anyone to apply. Spelling can be checked by a reference to a dictionary. A grammar book is required for ready reference. All employees are expected to observe as closely as possible these rules for maintaining accuracy.

*Reliability:*Reliable persons are those you can trust. They keep their word, do as promised, and tell the truth.

A delivery is made on the day promised. A telephone call is made at a time the customer is at home. If a customer complains, a reliable person follows through until the problem is solved.

A reliable person doesn't procrastinate. In contrast to this, unreliable clerks take the telephone number of the customer but never call. These procrastinators file a customer complaint to solve sometime tomorrow or next week.

If you have reliable workers, you can rely on them coming to work every day. You can rely on them to do an excellent job. Management can rely on them for professional or technical assistance and for help in a difficult situation.

Example: Reliability of power operations depends upon the reliability of employees and equipment. The following gives a measure of the reliability of power in four cities: [12]

1. Loveland (includes planned outages)
   99.977 percent; outage time equals 2 hours per customer per year
   Goal 99.99 percent; 53 minutes per customer per year
2. Fort Collins (excludes planned outages)
   99.986 percent; 73 minutes outage
3. Greeley
   99.997 percent; 15 minutes outage
4. Boulder (Public Service includes planned outages)
   99.99 percent; 53 minutes outage

In Loveland, shortages caused by animals were reduced from 50 in 1987 to 18 in 1988 by using fiberglass guards around key components.

Squirrels are the primary animal problem. This was a quality improvement.

Listening to the customer is a habit an employee has to cultivate. It is important to understand what the customer is trying to say.

There is a need to teach employees how to behave toward customer contacts—whether in person, by letter, or over the telephone.

Continued Course

Education on how to treat customers is a special course that should be given to every employee, and repeated on occasion. This course stresses courtesy, consideration, politeness, and accuracy. The emphasis is on the fact that a pleased customer always comes back. This means that the customer is waited on immediately, that there is little delay time, and that the customer is satisfied.

7.   Introduce Statistical Techniques

The statistical techniques that are most helpful in quality control include:

1. The time plot
2. Frequency distribution: f
3. The binomial count: np
4. The binomial proportion: p
5. The Poisson count: c
6. The $\bar{X}$ and R charts of Shewhart
7. Random time sampling

1.   *The time plot*—Figure 3.1 shows the trend of blood pressure measured on the right arm 20 times over 7 months. It shows 12 recent points falling between 120 and 129, showing a trend toward stabilization at a very acceptable level. This time plot gives a clearer picture of the long-term trend. This patient was on medication that brought the blood pressure under control.

2.   *Frequency distribution*—Figure 3.2 is a frequency distribution (called a histogram) that shows monthly use of electricity. The lowest figures were for daylight saving time months. The high figures were for the longest months with fewer daylight hours.

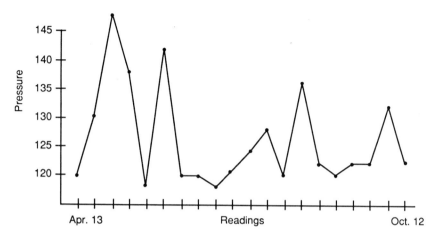

Figure 3.1
Blood pressure, April–October 1989.

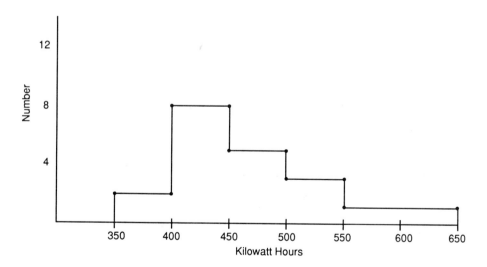

Figure 3.2
Histogram illustrating monthly household consumption of electricity.

3. *Binomial count np*—Figure 3.3 shows the binomial distribution, a "heads or tails" model:

1. The mean is $n\bar{p}$.
2. The variance is $\sqrt{n\bar{p}\bar{q}}$ or $\sqrt{n\bar{p}}$ as an approximation where $q = 1 - p$ $(q \rightarrow 1)$.

Example: Every work lot contains 300 documents. The number of errors found in 20 work lots are:

8	11	$n\bar{p} = \dfrac{89}{20} = 4.45$
4	6	
6	8	Control limits:
3	4	$4.45 \pm 33 \sqrt{4.45}$
2	6	$4.45 \pm 6.33 = 0$ and 10.78
5	5	
1	2	Upper control limit (UCL) = 10.78
0	4	Lower control limit (LCL) = 0
5	2	A count of 11 is out of control.
4	3	

4. *Binomial proportion $\bar{p}$*—Figure 3.4: The counts given in the above example can be converted to proportion p.

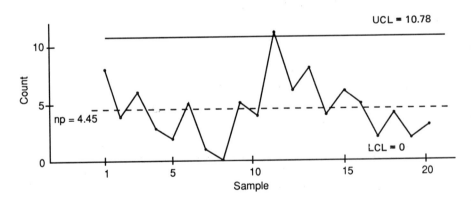

Figure 3.3
Binomial count $n\bar{p}$ *errors found in 20 lots of 300.*

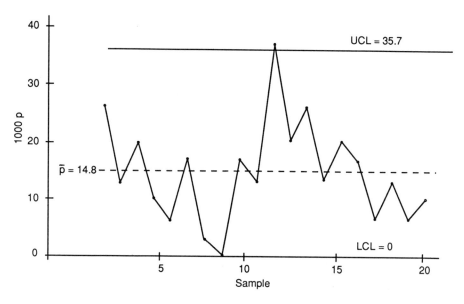

Figure 3.4
Binomial proportion p̄ errors found in 20 lots of 300.

Example: p

0.0267	0.0367	
0.0133	0.0200	$\bar{p} = \dfrac{89}{6000} = 0.0148$
0.0200	0.0267	
0.0100	0.0133	$\bar{p} = \dfrac{0.2968}{20} = 0.0148$
0.0067	0.0200	
0.0167	0.0167	
0.0033	0.0067	
0.0000	0.0133	
0.0167	0.0067	
0.0133	0.0100	

Sum   0.2968

Control limits:

$$\bar{p} \pm 3 \sqrt{\frac{\bar{p}\,(1 - \bar{p})}{n}}$$

$$0.0148 \pm 3 \sqrt{\frac{0.0148 \times 9852}{300}}$$

$$0.0148 \pm 0.0209 = 0 \text{ and } 0.0357$$

A value of 0.0367 is out of control; this is a count of 11.

5.   *Poisson c test*—Figure 3.5: For the Poisson distribution $\bar{x} = s^2 = c$

Control limits: $\bar{c} \pm 3 \sqrt{\bar{c}}$

A telephone customer had a number of false calls billed to him over a period of 10 months.

Number of false calls:

1	0
4	2
2	2
0	1
3	1
	16

$\bar{c} = \dfrac{16}{10} = 1.6$

Control limits:
$1.6 \pm 3 \sqrt{1.6} = 1.6 \pm 3.8$
$= 0 \text{ and } 5.4$

The values are in control and are to be expected unless the system is changed.

6.   *$\bar{x}$ and R charts*—Figures 3.6 and 3.7 illustrate measurements from a figure-eight timer; times were measured by a stopwatch.

Deviations are from 3 minutes in seconds (see Table 3.1).

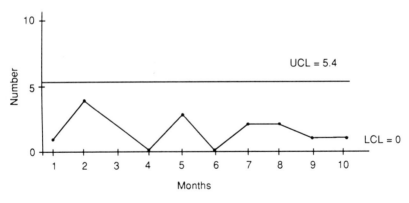

Figure 3.5
Number of false telephone calls. $\bar{c} = 16/10 = 1.6$; $c = 3 \sqrt{\bar{c}} = 1.6 \pm \sqrt{1.6}$; UCL = 5.4; LCL = 0.

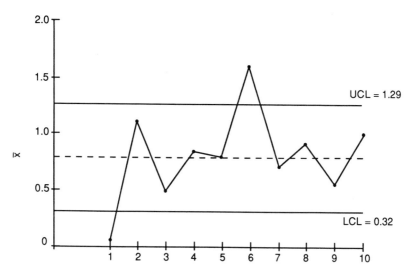

Figure 3.6
$\bar{X}$ *chart for three-minute timer (measured in 0.1 second).*

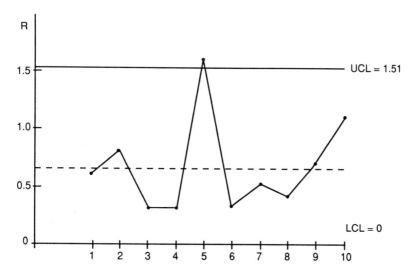

Figure 3.7
R *chart for three-minute timer (measured in 0.1 second).*

Table 3.1

Ten Replicates

Sample	1	2	3	4	5	6	7	8	9	10	Sum
1	−0.3	1.2	0.4	0.7	1.0	1.6	0.6	0.9	0.2	1.1	
2	−0.1	1.0	0.5	1.0	0.7	1.7	−0.8	1.0	0.5	1.5	
3	0.3	1.6	0.7	0.9	1.5	1.6	0.4	1.0	0.6	0.4	
4	0.3	0.8	0.4	0.8	−0.1	1.4	0.9	0.6	0.9	0.9	
Sum	0.2	4.6	2.0	3.4	3.1	6.3	2.7	3.5	2.2	3.9	31.90
$\bar{X}$	0.05	1.1	0.5	0.85	0.8	1.6	0.7	0.9	0.55	1.0	8.05
R	0.6	0.8	0.3	0.3	1.6	0.3	0.5	0.4	0.7	1.1	6.60

$$\bar{X} = \frac{31.90}{40} = 0.798 = \frac{\Sigma\bar{X}}{10} = 0.805 = 0.80$$

$$\bar{R} = \frac{6.60}{10} = 0.660$$

Control limits:

$\bar{X}$: $0.805 \pm 0.729 \times 0.660 =$ UCL 1.286   Sample 1 and 6 are out of control

LCL 0.324

R: $\bar{R} = 0.660$ Sample 5 is out of control

UCL: $2.282 \times 0.660 = 1.51$

LCL: $0 \times 0.660 = 0$

---

7.   *Random time sampling (RTS)*—Random time sampling is probability sampling applied to working time. The characteristics of RTS are as follows:

1. Population: 8 hours or 480 minutes daily, 5 days a week
2. Sample unit: a random minute
3. Sample: one or more observations of each person or object, using table of random minutes
4. Sample data: observe what each worker is doing at a random minute; obtain individual's cost per minute, as well as percentages
5. For the minute model, assume activity lasts at least one minute; apply cents per worker per minute to get costs; method of estimation is method used in the sample design

6. Samples can be from one to 16 on each person per day
7. Responses can be anonymous or made by an observer
8. Sample all employees

RTS can be used for several purposes:

1. To estimate what each worker is doing in percentages, e.g., whether the person is working, conversing, making telephone calls, or idle
2. To determine what the person is working on
3. To estimate the cost of production
4. To decompose the cost of joint activities
5. To estimate how much it costs to do each of several classes of activities
6. To estimate cost of idle and wasted time

We illustrate below how to estimate 1) a percentage and 2) the cost of an activity. We show how the data have to be collected.

Random time samples are best drawn by using Rosander's table of 4000 random minutes.[8,13]

Sample Designs and Calculations

Tables 3.2–3.5 show four sample designs with calculations.

Example: In a large data-processing division, federal government employees compared very favorably with private industry employees with regard to proportion of working time actually at work.

In one division of the IRS, 350 employees were 75-percent effective. That means 75 percent of the time, employees were working. The remaining 25 percent of employee time was accounted for by 1) annual leave, 2) paid sick leave, 3) no work assigned, 4) not working, and 5) other.

These data were derived from an RTS taken over a period of 26 months. Six studies made in private industry showed that 75-percent effectiveness was exceeded in four cases and less than 75 percent in two cases:[14]

Materials handling	82 percent
Nursing	80 percent
Engineering design	79 percent

Table 3.2
Proportion or Percentage of Time Spent on Activity A
(1 stands for doing activity A; blanks stand for doing something else)

Person no.	1	2	3	4	5	6	7	8	9	10	Sum
				Day no. (one random sample per day)							
1		1		1	1		1	1	1		6
2		1	1	1	1			1			5
3	1		1			1	1	1	1		6
4				1			1	1	1		3
5				1						1	4
Total x_i	1	2	2	4	3	1	3	4	3	1	24
p_i	0.20	0.40	0.40	0.80	0.60	0.20	0.60	0.80	0.60	0.20	4.80

$$\bar{p} = \frac{24}{50} = 0.48 \qquad \bar{p} = \frac{4.80}{10} = 0.48$$

$$s^2 = \frac{1}{10 \times 9} \sum_{1}^{10} (p_i - 0.48)^2 = 0.005511$$

$$s = 0.074$$

$$\bar{p} = 0.48 \pm 0.07$$

Clerical order, billing	76 percent
Above example	75 percent
Machine shop work	67 percent
Keypunch operation	52 percent

These figures do not include any money paid for sick leave or other leave. Including these adjustments would lower the six figures, and make the case with 75-percent effectiveness rank even higher.

Statistics and Deming's 14 Points

Three objections have been raised relative to the role of statistics in the 14 points:[15]

Table 3.3
Cost of Time Spent on Activity A

Person no.	\multicolumn{10}{c}{m (Random minutes)}

Person no.	1	2	3	4	5	6	7	8	9	10
1	20	20				20				0
2					18			18		0
3		30	30						30	0
4	15			15			15			0
5			12	12			12	12		0
x_i	35	50	42	27	18	20	27	30	30	0

$\Sigma x_i = \$2.79$ $M = 10 \times 480$ (10 days)
                          $m = 10$

Entries are costs in *cents per minute* for activity A.

Total cost $= \dfrac{M}{m} \sum x_i = \dfrac{10 \times 480}{10} \times 2.79 = \$1339.20$

Standard error

$\bar{x} = \dfrac{2.79}{10} = 0.279$

$s^2 = \dfrac{4800^2}{10 \times 9} \sum (x_i - 0.279)^2$ (x_i in dollars)

$s = 209$

Cost of A: $\$1339 \pm 209$.

1. A very different statistician is needed.
2. The 14 points have very little to do with statistics.
3. Statisticians cannot transform Western management.

   1. *A new kind of statistician is needed.* There are some statisticians who remain very close to the standards set by Deming. They are familiar with the application of the above seven techniques to all kinds of quality problems.

Table 3.4
Cost of Activity A for $n_i = 2$

Person no.	Cents per minute per day				
	1	2	3	4	5
1	20	0	0	0	20
	0	20	0	20	20
2	0	0	15	15	0
	15	0	15	0	0
3	30	0	30	30	0
	0	30	30	0	30
4	12	0	12	12	12
	0	0	0	0	12
5	0	20	0	20	0
	20	20	0	0	20
x_i	97	90	102	97	114 500
					5.00

Entries are cents per random minute.

Population = $5 \times 480 = 2400$. $\Sigma\, x_i = 500$ cents = \$5.00; $\bar{X} = \dfrac{5.00}{5} = 1.00$.

Total cost = $\dfrac{5 \times 480}{5} \times 5.00 = \$2400$.

$$s^2 = \dfrac{2400^2}{5 \times 4} \, \Sigma\, (x_i - 1.00)^2 \quad (x_i \text{ in dollars})$$

$s = 96$

Cost of A = $\$2400 \pm 96$

Table 3.5
Replicated Sampling for A Percentage and Cost

Person no.	Replicates for one day				Sample Frequency of activity A	
	1	2	3	4		
1	345 P.M.	912 A.M.	948 A.M.	133 P.M.	2	4
2	330	204	217	834	3	4
3	858	1150	141	123	1	4
4	854	1023	1049	317	4	4
5	440	834	1124	1053	2	4
6	924	1117	111	254	1	4
Sum	4	5	1	3	13	24

Summary of Table 3.5

Person no.	n	Activity A	Proportion	Cost
1	4	2	0.50	$M = 480$; $m = 4$
2	4	3	0.75	
3	4	1	0.25	$M = 120$
				m
4	4	4	1.00	15 cents per minute is average
5	4	2	0.50	
6	4	1	0.25	
Sum		13	3.25	

Proportion

$$\bar{p} = \frac{13}{24} = 0.54 \qquad C = 120 \times 13 \times \$0.15$$

$$\bar{p} = \frac{3.25}{6} = 0.54 \qquad C = \$234$$

$n\bar{p}$ Test

$$n = 4 \quad n\bar{p} = \frac{13}{6} = 2.22$$

Limits: $2.22 \pm 3 \sqrt{2.22} = $ UCL $= 6.65$

$$\text{LCL} = 0$$

c Test

$$\bar{c} = \frac{13}{4} = 3.25$$

Limits $= 3.25 \pm 3 \sqrt{3.25} = 8.7$ and 0

Variance test

$$s^2 = \frac{1}{4 \times 3} \sum_{1}^{4} (x_i - 3.25)^2 \qquad \begin{array}{l} s = 0.73 \\ 3s = 2.19 \end{array}$$

Limits $3.25 \pm 2.19 = 5.44$ and 1.06.

All points fall within limits except 1.06. The cost of activity A for this day was $234.00.

The statistician needed has an M.A. degree in technical statistics and abilities to solve real-world problems. This statistician: 1) helps management solve statistical problems anywhere in the organization and 2) stresses management problems, not statistical problems of the statistician.

The statistician's job is to solve problems that have statistical aspects. Only the statistician is capable of finding these aspects and providing an answer to them. This requires a broad-gauged statistician who is problem-oriented.

2.   *The 14 points have very little to do with statistics.* In his *Quality Productivity and Competitive Position,* Deming uses about 45 statistical problems and charts to develop the text. He applies the following:

1. Binomial counts: $\bar{x} = np$
   Control limits: $np \pm 3 \sqrt{npq}$
2. Binomial proportion p: $\bar{x} = \bar{p}$
   Control limits: $p \pm 3 \sqrt{\dfrac{pq}{n}}$
3. Poisson count c: $\bar{x} = s^2 = c = \sum \dfrac{x}{n}$
   Control limits: $c \pm 3 \sqrt{c}$
4. Square root approximation of c
   $(\sqrt{x_1} + \sqrt{x_2} + \ldots \sqrt{x_n})/n = \sqrt{\bar{x}_{avg}} \pm 1.5$
5. $\bar{x}$ and R charts
   $\bar{x}: \bar{x} \pm A_2 R \qquad A_2 = 0.729$ for $n = 4$
   $\qquad\qquad\qquad\qquad = 0.577$ for $n = 5$
   R: UCL $= D_4 \bar{R}$
   $\qquad$ LCL $= D_3 \bar{R} \quad D_3 = 0, D_4 = 2.282$ for $n = 4$
   $\qquad\qquad\qquad\qquad D_3 = 0, D_4 = 2.115$ for $n = 5$
6. Approximation of $s =$ std. dev.
   $s = \dfrac{\bar{R}}{d_2} \qquad n = 4: d_2 = 2.059$
   $\qquad\qquad n = 5: d_2 = 2.326$
7. Line charts
8. Linear regression

This analysis shows that statistics has a lot to do with the 14 points. Statistics is necessary but not sufficient.

3.   *Statisticians cannot transform Western management.* Alone they cannot do this, but they can help. They are not top managers but they can communicate their mission to management.

8. Collect Data

Methods

Decisions are no better than the data on which they are based. Hence the emphasis is on collecting high-quality data. The real question is: How will the sample data be used? Data are collected in a number of different ways.

- 100-percent coverage (a census)
- Sampling

  Simple random
  Replicated
  Stratified random
  Random time
  Discovery
  Systematic

- Questionnaire
- Interview
- Observation
- Data sheet
- Sample in order of production
- Experimental design

The basic procedure to follow is to:

1. Mail out documents to everyone in the sample.
2. Follow up by mailing two reminders to nonrespondents.
3. *Interview all* the remaining nonrespondents.

Hints for Collecting Data with a Census

1. A census requires careful planning.
2. Be sure to get 100 percent.
3. Test your questionnaire or data sheet for flaws.
4. Train the interviewers.

Sample

1. Design a sample to meet stated purpose.

2. Mail out sample; follow up by interviewing nonrespondents.
3. Stratify if population consists of different groups; do not stratify for percentages.
4. Get all sample elements—do *not* settle for 90 or 95 percent.
5. Divide sample so 100 percent of customers are included every six months.
6. Analyze sample differences for real differences.
7. Sample in order of production for quality-control charts.

Questionnaire

1. Design to meet stated purpose.
2. Avoid words that load the questionnaire—leading questions.
3. Use neutral words.
4. Test questionnaire on those who will answer it.
5. Selected unbiased interviewers.
6. Include identification question if necessary.

Observation

1. Select objective observers.
2. Instruct observers to record what they see and hear.
3. Record what bears directly on purpose.

Data Sheet

1. Design to meet the purpose.
2. Include instructions.
3. Include identification.
4. Test data sheet to de-bug.

A simple trend graph is a very useful device. It is plotted on ordinary graph paper with time measured along the horizontal axis.

Interviewers

Interviewers need to be trained in the content of, and the answers to, questions that are asked. The form of the questions needs to be discussed. Changing one word can change the meaning. Test the final questions for clarity, to eliminate ambiguity. Just any set of questions about the subject matter will not do. (See Stanley Payne, *The Art of Asking Questions.*[16])

Possible answers to questions need to be studied for relevance and irrelevance.

Experiments

Experiments have to be carefully designed using randomization, replication, and balancing. The following illustrate some ways to experiment:

- A simple experiment: How does y change with changes in x?
- Paired comparison
- Paired x and y—comparing two samples treated by several laboratories
- Random blocks

Example: The press reported the results of tests made by the Environmental Protection Agency to determine if differences in plutonium existed between Rocky Flats near Denver and a test site in Nevada.[17] The level of plutonium in lung tissue was 5.1 pc/kg (picocuries per kilogram) in a herd of 10 cattle at Rocky Flats. In a herd of 10 cattle at the test site in Nevada, this value was 1.34 pc/kg. The level of plutonium at Rocky Flats was about four times that in Nevada. The variation among animals within the herd at each site was very high, showing that the averages were unstable and in another test could easily be reversed. This is what happened; a second test on a herd of eight animals in Nevada gave a value of 12 pc/kg, more than twice the figure at Rocky Flats. The quantity to test was the difference between 5.1 and 1.34 or 3.76, not the ratio of the two values. The difference was of no significance due to the high within-herd variation.

The amount of plutonium in cattle lymph nodes was slightly higher at Rocky Flats (5 pc/kg) than at Nevada (4.5 pc/kg), a difference of 0.5 pc/kg. Again this difference was of no significance because of the wide variation among animals within herds—a reported range from 0.7 to 60 pc/kg.

Certain experimental techniques are standard for this type of test. Animals should be assigned at random to the two sites to eliminate any inherent differences in the reactions of the animals to plutonium. Animals to be compared should have been exposed to plutonium for the same length of time. These tests should be repeated to obtain limits to the ranges of averages and variations. Equal numbers of animals at the two sites should be used; they should be paired if possible.

This test was reported in the press as showing a level of plutonium four times as great at Rocky Flats as at Nevada: 5.1 compared to 1.34. This was another example of apple arithmetic: 6 apples ÷ 3 apples = 2 apples.

Improving Data Collection

Data collection requires *legible writing* to avoid errors. Such errors include the following:

- Decimal point in wrong position
- 1 and 7 confused
- 4 is taken to be a 9
- Illegible writing
- Numbers not arranged in vertical columns
- *a* and *o* confused
- *e* and *i* confused
- 8, 3, and 5 confused

Data collection requires:

1. A well-defined problem or an exploration of some situation
2. Clear, legible data
3. A correct analysis aimed at solving or uncovering a problem
4. Taking action as a result of the analysis
5. Follow-up on action taken to see if it is appropriate
6. Making changes to improve data collection and analysis

Measuring User Satisfaction

Whether a customer is pleased, satisfied, dissatisfied, or lost can be determined by continual surveys of customers. A company doesn't know how customers appraise services if it makes no attempt to find out. The basic data should come from continual customer surveys as described above.

Outspoken customer complaints are easy to obtain, but they are only the tip of the iceberg. They represent a biased and incomplete sample.

Cautions in Using Pareto Analysis in Services

Pareto analysis is based on an analysis of the data into the causes that are "the trivial many, the vital few." There are several objections to this idea applied to both products and services:

1. Percentage of occurrence does not measure the importance or significance of the error or defect.
2. There is a need to drive all errors to zero regardless of the frequency of occurrence.
3. The "trivial" error may occur only once, but it can be dangerous, if not fatal.
4. If our goal is quality improvement, we must aim at getting rid of defects and errors regardless of their frequency of occurrence.
5. It may be difficult to determine which errors or defects are more important than others.
6. "Importance" of errors and defects may not be an important factor.

Faulty Sampling in Public Opinion Polls

The following data show how opinion polls may go awry and pick the wrong candidate.[18]

New Hampshire Primary, February 1984

Candidate	ABC/Washington Post poll	Cable News poll	Actual Vote
Mondale	32%	38%	29%
Hart	25%	22%	40%
Sample size	450	500	
Published margin of error	5%	4.4%	

The false sampling errors published were 5 percent and 4.4 percent.
    Calculation of errors published:

For $n = 450$ $s = \sqrt{\dfrac{1/2 \times 1/2}{450}}$ $2s = 0.0472$ or 5%

For $n = 500$ $s = \sqrt{\dfrac{1/2 \times 1/2}{500}}$ $2s = 0.0448$ or 4.4%

The sampling requirements were not met, so "margins of error" are meaningless. Requirements to be met in such polls are:

1. A random sample of eligible voters who are going to vote
2. Not conducting a telephone survey that invalidates the sample, by ignoring:

   - Those not at home or whose line is busy
   - Ineligibles: those who listen but give no response
   - Those who won't answer
   - Nonresponses and those in doubt

The "margin of error" figures are wrong and useless.

The two polls showed Mondale winning, but Hart was the real winner, receiving 15–18 percent more votes than the polls had indicated. Hart's 40 percent and Mondale's 29 percent showed the tremendous amount of bias in both polls. The margins of error were falsely based on biased "samples" of 450 and 500.

Since these were not random samples selected from all eligible voters, the margins of error do not apply. They are null and void, without any meaning.

These pollsters applied probability sampling without having the least understanding of what conditions had to be met.

9.   Emphasize Teamwork and Responsibility in Problem Solving

Original ideas come from individuals, not from groups or masses. Look for the person who thinks independently, who is innovative, and who suggests improvements and new products and services. Look for the self-starter who originates improvements on his or her own initiative. We cannot rely on consensus for quality improvements. These arise, most of the time, from alert and knowledgeable individuals. The goal is to stimulate this attitude in most employees.

Brainstorming is fine when the company wants a snappy slogan or something that doesn't require technical knowledge. However, it can be a waste of time unless participants in the group have relevant knowledge. This means that every individual in the group should know something about the problem.

When technical issues are involved, care should be taken so that brainstorming does not degenerate into a session in which members only share ignorance. The author was a member of such a group.

Teamwork, or the use of a task force, does not require brainstorming. Every member of the task force contributes to the solution of the problem by applying his or her special knowledge—the accountants, lawyers, statisticians, economists, marketing people, transport specialists, etc.

Teamwork requires the cooperation of several members of a task force. An example of a task force in transportation:

Two lawyers
Two operations research specialists
Two accountants
Two statisticians
One economist
One rail traffic expert

The purpose of this study was to measure the shortage of freightcars in the United States using a nationwide sample study. This was the first time this study had ever been made. The members of the task force had these responsibilities:

- Lawyer—direct, manage
- Statistician—design, control, and interpret sample data
- Accountant—receive and edit sample data sheets received by mail
- Operation research—tabulate sample data
- Traffic expert—solve rail traffic problems
- Economist—serve as observer

Correcting Faults from the Top

In theory, top-level management is responsible for the quality of all products and services that the company offers for sale. In practice, this is impossible. In services it is the people behind the counter, behind the desk, or waiting around who determine the quality of service.

Top-level management is responsible for company-oriented activities such as:

- Finances: payroll and profits
- Policies
- Hiring of top-level employees
- Budget
- Training

- Travel
- Sales
- Taxes
- Purchase and repair of equipment
- Purchase and repair of machinery
- Purchase and modification of the computer system
- A company-wide probability sample study
- Arrangement of goods in a retail store
- Safety training
- Safety practices

Correcting System Faults at Lower Levels

The first-line supervisor, administrative assistant, or secretary may correct system quality defects. These defects include failure of heating, air conditioning, lighting, telephone, and equipment such as typewriters, desk calculators, and the computer. In these and other situations, the appropriate people are called at once. The trouble is never reported above the level at which it occurs unless it persists, or unless it is difficult to obtain the necessary service. Many faults can be corrected by the supervisor.

Certain types of troubles, corrections, or changes are reported higher up, including those involving large financial outlays, unbudgeted items, drastic changes in processes or procedures (sampling for 100 percent tabulation),and drastic changes in personnel, space, equipment, machinery, travel, and training.

A system correction procedure manual is needed to define where responsibility lies for taking corrective actions for various situations. The manual needs to be developed and used in system correction programs and for training.

Correcting Special Faults

Special faults include errors and defective products made by the employee that he or she can correct, such as getting articles interchanged, ignoring the customer, or showing attitudes and behavior toward the customer that can be corrected.

Zero Errors

The most important defects and errors to concentrate on are those that should be *zero*. Examples include:

- An airplane taking off with ice on the wings
- Substituting carbon dioxide for oxygen or argon for oxygen
- Substituting a fatal drug for a safe drug
- Making errors in billing
- Paying ineligibles and not paying eligibles
- Making errors in payroll
- Making errors in typing and word processing
- Making errors on insurance claims
- Bank tellers' errors
- Retail store clerks' errors
- Using excessive time to perform an activity
- Shoring on construction projects
- Making errors in nuclear power plants

This means a division of responsibility between management and employees of at least 50/50 in transportation, health, construction, and nuclear power plants. It may mean 100-percent responsibility of employees for some errors, idle time, and many other kinds of wasted time.

Correcting Chronic Difficulty

A quality-control chart corrects chronic difficulty. Two actual cases will be cited. The charts in these examples are supposed to correct only special causes, but they correct chronic causes instead.

Case 1: A canning factory sold 12-ounce cans of sauerkraut, but quality-control charts showed their average weight was about 14 or 15 ounces. The canning method was changed to obtain a weight closer to 12 ounces. The factory finally obtained a range between 12 and 12.5 ounces. This was a chronic cause, not a special cause, and it was corrected by using a quality-control chart. This was a breakthrough.

Case 2: The quality-control chart was used to correct a similar problem in a steel mill. Steel ingots that were supposed to weigh 5200 pounds actually weighed an average of 5400 pounds. A statistical quality-control chart revealed this excess weight. A change in the process brought the average down to about 5250 pounds, resulting in an annual savings of $175,000. A quality-control chart revealed the excessive weight and corrected it. This was also a breakthrough.

Special Causes

Common causes exclude special causes. Special causes are what an individual employee can avoid and correct:

- Errors, mistakes, blunders
- Delays
- Lost time
- Wasted time
- Safety violations
- Ill-mannered behavior

The individual is responsible for these non-quality actions. The goal is to prevent them and increase productivity. A continuous course in how to prevent non-quality actions is required.

CHAPTER 4

Quality Improvement

This chapter continues the outline of a quality program begun in Chapter 3. The emphasis here is on identifying, solving, and preventing problems.

10. Detect and Prevent Errors

The nation is plagued by errors, mistakes, and blunders. Examples are:

- Airplane pilot error
- Truck driver error
- Schoolbus driver error
- City bus driver error
- Hospital errors by doctors and nurses: mix-up of drugs and gases
- Nursing home errors
- Retail store errors
- Ship's captain error
- False charges on telephone bills
- Construction failures
- Computer errors
- Arrogant employee errors

These are just a sample of all the errors made. Several specific cases and the fatal results of some of these errors are described below.

Errors

1.   A steady customer was charged $166 in error on monthly billing for a purchase of clothing. The bills came from the central billing office of a large chain store. The customer kept a record of all her purchases using a credit card; this is how she discovered the error.

The billing office claimed she had loaned the credit card to a friend or relative, or forgotten about the purchase. The company insisted to the end that the customer was guilty of the error.

A study of the sales slip showed that the purchase was made on a Saturday at Store 6. The customer never shopped at Store 6 and never on a Saturday. The signature on the sales slip was not the customer's.

This should have settled the case. It did not. An affidavit showing that the card was used illegally was sent to the customer. No apology was sent; the company kept quiet. One of several circumstances could have caused this error:

- The clerk at the store could have copied the 10-digit credit card number and made the purchase.
- The number could have been given to a friend.
- The central credit bureau might have wanted to avoid taking a loss, so the bureau shifted it, in this case, onto the wrong customer.

The store never apologized or disclosed what happened. They were afraid of errors. No Nordstrom treatment here! They followed the practice that the less said to the customer about errors, the better.

2.   A man suffering from emphysema broke a bone and was taken from a nursing home to a hospital. He was put on oxygen but died within 15 minutes. Why? The cylinder brought in from a dim storage room contained carbon dioxide, not oxygen. Who was to blame? The individual? Management? The system?

- Was management to blame for hiring the individual, or for not training him or her?
- Should the lookalike cylinders have been stored together?
- Should the cylinders have been conspicuously labeled?
- Should the giving of the gas have been carefully monitored?
- Should a test have been made to ensure the gas was oxygen?

- Do the individuals have any obligation relative to storage, labeling, identifying, and administering?
- Is the employee as much to blame as the system or management?

Epilogue: The coroner ruled that the man, who was over 90 years old, died of natural causes! As a result of this error, a training class was started.

3.　Similar cases include these:

- Argon connected to oxygen line: two dead
- Nitrogen mixed with oxygen: one dead
- Bottles mixed; cocaine for phenobarbital: one dead

There was also a case in which the oxygen valves on two similar anesthesia machines stuck, resulting in two deaths. This latter example raises the question as to the medical personnel's obligation to test that a machine is working properly. In this case the two machines had undergone a maintenance checkup a few months before.

4.　Airplane crashes: An airplane landed at National Airport, Washington, D.C., during a snow–sleet storm. It was on the ground being de-iced for at least 45 minutes, yet the pilot of another plane reported he saw ice on its wings. The plane took off and crashed on the 14th Street bridge. The casualties included several persons driving home from work as well as 70 passengers in the plane. Who is to blame? The system? Management? Individuals?

A plane took off from the New Orleans airport despite reports of wind shear in the area. It crashed into a residential area. Who is to blame? The system? Management? Individuals?

Human errors of employees were responsible for these tragedies. What can be done to prevent them? Are employees told that zero errors are a "must"? Are they told this when they are hired? Are employees taught error-prevention programs periodically? The answer should be "yes," because drastic measures are required when lives are at sake. This is not the usual factory situation.

5.　In other transportation methods, too, serious accidents are caused by employee errors.

For instance, a train engineer had to check at a certain point if another train was coming his way on the same track. If so, he was to shift to another track. The engineer read a notice in error, reading "yesterday" as "today." As a result, his freight train went onto a collision course with another train.

The collision occurred under an overpass on a road north of Denver. As a result:

1. Five people were killed.
2. Millions of dollars of damage was done to the railroad track and rolling stock.
3. The overpass was blown to bits.
4. Automobile traffic was stopped in both directions.

Repair was required for the highway, the track, the freightcars, the locomotives, the overpass, and the road bypass.

This shows the chain reaction a simple error creates. No steps were taken at the time to prevent another similar mistake.

6.   Nuclear power plant: It was reported in the press that a workman in a nuclear power plant, while doing cleanup work, somehow got his clothing caught in a control switch. He gave it a yank and closed down the entire plant. It was not explained how a situation such as this arose, or how it could be prevented.

The nuclear power plants at Three Mile Island and Chernobyl underwent other, more serious disasters. While Chernobyl was much more dangerous, both accidents were caused by workers and management. Lack of training and ignorance of how the plant operated and how the safety system worked led to the disasters.

Delays

Another kind of error is causing or failing to prevent delay. There are numerous kinds of delays. In an ophthalmologist's office, delays might be as follows:

	Waiting	Service
Waiting in waiting room	15 minutes	
Eye test		5 minutes
Waiting	15 minutes	
Glaucoma test		5 minutes
Waiting	10 minutes	
Doctor's service		5 minutes
Total	40 minutes	15 minutes

In a data-processing unit, a random time sample was carried out for five consecutive days. Four units reported the *delay in receiving work* during a 40-hour week:

Unit 1: 9.0 percent
Unit 2: 5.9 percent
Unit 3: 4.2 percent
Unit 4: 7.2 percent

All these employees were idle while waiting for work. Immediate steps were taken to correct this situation, but their effect is unknown.

In a large retail store, a customer ordered a set of drapes. After the promised delivery date, the customer made several telephone calls. Still no goods were received. Finally, the customer drove the 10 miles to the store, and checked into the matter:

1. The store clerk could not find its sales slip, so the clerk used the customer's!
2. After considerable searching, the drapes were found hidden away in a cubbyhole.
3. It was estimated that the drapes had been there at least 10 days.

Wasted Time

The cost of wasted time in services can be terrific. It takes many forms:

- Idling on the job
- Making personal telephone calls
- Taking long lunch hours
- Ignoring customers
- Not showing up for work (no explanation given)
- Enduring delays in receiving critical shipments
- Being unable to alphabetize a person's name
- Visiting with other employees
- Making a customer wait

Anti-Quality Behavior and Attitudes

Anti-quality behavior and attitudes include carelessness, indifference, ignoring customers, finishing a job before waiting on customers, putting a telephone call ahead of a customer already being waited on, promising more than the product or service can deliver, charging insurance premiums for a condition that does not exist, e.g., complete loss of income. Anti-quality attitudes among managers and employees are demonstrated by:

- Rudeness
- Carelessness
- Indifference
- Taking customers in the wrong order
- Showing dislike for certain customers
- Failing to keep promises
- Ignoring customers

An example of carelessness was an operator who punched two 6's instead of one, entering $166,000,000 instead of $16,000,000, an error of $150,000,000. This occurred in a property-appraisal office in Denver.

Defective Products Purchased

Every service organization must purchase products. Some of these are defective, as in the following:

- Medical—a defective anesthesia machine; carbon dioxide mistaken for oxygen; argon mistaken for oxygen.
- Transportation—Of 30 new rapid-transit buses, 15 are defective. The 15 are not returned to the factory; instead, the defects are corrected by the transit company.
- Banking—checks printed with wrong numbers.
- Government—an unfit computer system; an individual house water meter that does not work.
- Retail trade:

  A travel agency sells an airplane ticket that calls for meeting a bus, with no instructions as to where to find the bus: The airplane is in Denver; the bus is in Seattle.
  The case of drapes mentioned above; 10-day delay within store.
  A hotel room whose heating system doesn't work in the winter or whose air conditioning doesn't work in the summer.

We have our work cut out for us: to get rid of errors and defects, mistakes, and blunders.

Unsafe Methods and Actions

We have already mentioned some unsafe methods and actions:

- Ice on wings of an airplane
- Poor brakes on a schoolbus
- Mix-up of medicine and drugs
- Mix-up of gases in a hospital
- Lack of shoring on a construction project
- Truck driving too fast on a curved road
- Defective anesthesia machine in a hospital
- More than one train using a single track

These errors are more common in the areas of transportation, health services, construction, pharmaceuticals, and nuclear power plants.

Retail Stores

A survey by Arthur Young and Co. of retail stores found that companies lost $1.8 billion in 1987 through shoplifting, theft by employees or vendors, and errors in paperwork.[19] This compared with $1.5 billion in losses during 1986.

The study included 119 retail companies operating in 47,000 locations with total sales of $127.5 billion. The losses were divided as follows:

Discount stores	$1.2 billion
Department stores	$0.364 billion
Specialty shops	$0.233 billion

From 1982 through 1987, shoplifting increased as follows:

Discount stores	1.5–2.1%
Department stores	2.0–2.1%
Specialty shops	1.7–2.2%

At the top of the list of shoplifted items was ladies' designer apparel. At the bottom were jewelry, shoes, and men's clothing.

The number of suppliers raises several questions about quality and price (Deming's point 4):

- One supplier may not carry all the products required.
- Several suppliers are necessary to obtain the variety of products needed.
- One supplier is not enough, except for some firms.
- The distance to suppliers makes a difference; distance doesn't over-rule quality.
- Suppliers must have statistical quality-control programs that will show the buyer that quality is built into every order. The suppliers must submit evidence to prove it.
- Purchases are based on quality of the product, not on the lowest price.

Hospitals and Nursing Homes

Three measures of hospital quality control are confused:

1. Institutional quality

    Vacant beds
    Admissions
    Discharges
    Number of patients who survive surgery
    Ratio of nurses to doctors
    Percentage of beds used

2. Quality performance of professionals and other personnel

    Doctors
    Nurses
    Clerks
    Attendants
    Assistants
    Volunteers

3. Patient service quality

Patient service quality is what medical services are for: to serve the patient. All the rest are supportive. The quality of medical service rendered to the patient is vital and should receive top priority.

Safety is the primary quality characteristic in health services, transportation, power plants (including nuclear), construction, and certain chemical plants. The truth of this statement is illustrated by numerous examples described later.

There is a need to drive out the fears of employees, customers, and patients (Deming's point 8).

In a nursing home the problem is not driving out employee's fear, but patient's fears. These fears are associated with administrators, employees, nurses, supervisors, thieves, rapists, insane patients, fire, and lack of money.

Examples

1.   A rancher in a nursing home sold his ranch and cattle for $100,000. He was running out of money and asked: What am I going to do when it is all gone?

In the nursing home, he lost his wallet, his shaving outfit, and some of his clothes to a thief. He saw two other patients wearing his clothes. He asked for a lock on his door but was refused.

2.   A woman in a nursing home had the following personal property stolen: three nightgowns, a new wig, jewelry, a sweater, and some food. When her daughter let the nurses and manager know that the new wig had been stolen, it suddenly appeared. This is an example of stealing by employees.

3.   Two owners of a nursing home were charged in court with stealing $14,000, the entire trust fund of the patients. This fund contained cash belonging to the patients and was held by management for safekeeping. The patients, it was assumed, were incompetent to have any money! Some are.

4.   A woman who was mentally normal but physically weak was moved into a room with another patient. Her new roommate turned out to be insane: This can happen when mentally ill persons are allowed to enter nursing homes. The normal woman became very frightened and told her daughter, who complained to management. Management moved the insane woman to another room.

Observations in several nursing homes reveal that patients will complain to outsiders but not to an employee, a nurse, or management. They are afraid to complain to anyone connected with the nursing home.

5.   Hospitals make errors. The following are on record:

A cylinder of carbon dioxide was mistaken for a cylinder of oxygen, resulting in one death.

A supplier mistook cylinders of argon for cylinders of oxygen, resulting in two deaths.

A nurse in a hospital emergency room mistakenly gave cocaine instead of phenobarbital, resulting in one death.

Two anesthesia machines of the same model were used during two successive months. Each has a defective oxygen valve, resulting in one death on each machine.

Two women scheduled for surgery were not clearly identified. Their surgical operations are interchanged. No deaths resulted since the operations did not involve vital organs.

Who is responsible for these errors?

Was management to blame for hiring the individual?
Should the lookalike cylinders be stored together?
Should the cylinders be conspicuously labeled?
Should administration of the gas have been carefully monitored?
Should a test be made to show the gas is oxygen?
Do the individuals have any obligation relative to storage, labeling, identification, and administration?
Who is to blame—employee, management, the system?

6.   Transportation is subject to errors and failures. Passengers want to know:

Safety record
Maintenance practices
Experience of the driver, pilot, or engineer
Condition of the airplane, train, bus, boat
Operating environment: danger of ice on wings, capability of driver, weather conditions

Additional Examples of Errors

1. This is a case in which one department—the computer department—questioned the soundness of a nationwide probability sample study designed by the statistics department. A detailed set of specifications had already been worked out by the latter department.

A meeting was called with the National Bureau of Standards (now the National Institute for Standards and Technology) to review the plan. No one in the statistics department knew anything about what was going on until the meeting, when they suddenly discovered its purpose.

The plan was discussed and handed out for review. The NBS's representative concluded that the detailed specifications were among the best he had ever seen. This settled the question, but not the conflict. The secret attempt failed to discredit the statisticians.

2.   Another case involved an entire agency's attack on the probability sample work of the statistics division. A meeting of the entire agency staff was arranged. Again the statistics division was kept in the dark; only the chief statistician was asked to attend the meeting.

At this meeting, a sampling expert from the Bureau of the Census was the center of attention. At least 10 high-level officials attended. There was a lengthy discussion between the Census expert and some officials.

Finally the question arose whether the Census Bureau should monitor the sampling work of the statistics division. The Census expert objected. He argued that the chief statistician was quite capable of monitoring the sample.

The meeting concluded with a decision that the chief statistician should make weekly reports to the Census Bureau.

The Census expert didn't like this idea, and after two or three reports, the entire matter was dropped. No one objected. The agency-wide attack failed because the statistics division had plenty of evidence to show that its work was successful and directly applicable to the agency's purpose.

An Error-prevention Program

The goal of an error-prevention program is to get rid of all kinds of errors. The goal is *zero errors*. A few of the wide variety of errors to eliminate are:

- Billing errors
- Claims errors
- Nuclear power errors
- Medical errors
- Transportation errors (airplane, bus, train)
- Construction errors
- Premium errors
- Retail store errors
- Hospital errors

- Nursing home errors
- Sales slip errors
- Bank transaction errors
- Real estate transaction errors

An error-prevention course should periodically call attention to the sources, or potential sources, of error and their prevention. From the foregoing list, it is easy to find plenty of content material for such a course.

In medicine, error-prevention instructions could include:

- Use easily read identifications.
- Store dangerous medicines in separate rooms or cabinets.
- Test to see if the drug or gas to be administered is correct.
- Identify patients at all times.
- Use drugs according to doctor's orders: type, frequency, and amount.

Examples of How to Prevent Errors

1. A household water meter had given a previous reading of 4900 gallons. The present reading was 300 gallons. The utility assumed the meter had read to its limit of 1 million gallons and then began to read from zero. The meter did not work over a long period of time. No one had informed the billing office that the meter had failed at 4900 gallons and was reset to zero. Meter readers lacked communication with the billing department. The bill read as follows:

Previous reading	Present reading	Gallons used	Monthly bill
4900	300	995,400	$1,346.19

The figures shown above were calculated as follows:

1,000,000	995,100
$-4,900$	$+300$
995,100	995,400

The usual bill was $18 for 12,000 gallons; this bill was for May and included extra water for the lawn. The error grew out of lack of communication between billing and maintenance departments in the city government.

2.   An example of an excessive electric bill for July comes from Tampa, Florida. The bill was for $5,062,599.57. The utility discovered that the family owed $146.76. The mistake was made by a keypunch operator who entered a large transaction number in the computer on the line where the dollar amount should have gone. Where is verification of keypunch operators?

3.   Not everybody can count. A salesperson counted nine articles and entered the number as five. A clerk couldn't get the right count for about 350 manila folders in a file. Another clerk had a hard time counting 300 documents. One solution is count documents in groups of 10, and then add the groups.

4.   An erroneous formula was programmed into the computer, although the correct formula was available from a consultant. This resulted in about 26,000 errors being read out and published. A computer programmer needs assistance with advanced statistical formulas.

5.   A computer librarian may make a serious error. A nationwide sample usually ran about 230,000 documents based on manual control. At the end of the year, the computer gave a count of 212,000 documents. A thorough check showed that four reels of tape had been omitted, which amounted to 17,000 documents. This checked with the manual system. Obviously, the librarian needed closer control over tapes for a given job.

6.   A wrong statistical formula was used: $s^2 = pq/n$. This formula is used only when $p = x/n$, not when $p = x/y$. The regular normal distribution is used when sigma σ is assumed known but a t distribution should be used when only s is known.

7.   Errors are made on the basis of judgment in planning. A nationwide sample was designed to give data on an annual basis. This design was in accordance with the plans of the executive committee. When the study was completed, top officials criticized it because it was not run on a monthly basis.

The study was not run on a monthly basis because the sample would have cost too much. Top planners did not agree. A plan that was satisfactory one year proved undesirable the next year. Planning failed at the top.

A similar problem arose in an entirely different agency. Instead of sample data covering the entire United States, for a year, top officials wanted the data by state. Planning again failed at the top.

8.   A national weekly reported how modern psychology had helped

a hosiery manufacturer solve a serious problem.[20] According to the report, the manager had reason to believe that employees annually stole a million dollars' worth of nylon hose. He called in detectives, but they discovered nothing.

Then the manager called in a firm of consulting psychologists. They came with cameras and unidirectional screens. They too found nothing. They were about to leave when one of the psychologists asked the manager the question that broke the case: How had he arrived at a million dollars' worth of stolen hosiery?

The manager had had one of the best operators run a test to determine the hose output per unit of yarn input. A test of one operator was thought sufficient, since all operators were using semiautomatic machines. From the test the amount of yarn per pair of hose was calculated, and this figure was divided into total yarn input for the year to obtain expected production. This figure was one million dollars higher than production for the year; hence the suspicion of theft.

The explanation of what had happened was simple. The amount of yarn required to make a pair of hose was lower for the test operator than it was for all operators. The millions dollars' worth of hose was not stolen because it was never produced! The million dollars was a bias in the estimate of the output. It was also the extra cost the manager had to pay because not all the operators had the productivity of the test operator.

Three conclusions can be drawn from the record: 1) this was *not* a legal problem; 2) it was *not* a psychological problem, if we exclude the manager himself; 3) it was a statistical problem in which an estimate was made from a biased sample of one. This problem was not solved by modern psychology but by the use of modern statistics. Application of one or more of three statistical techniques would have revealed the true situation: 1) a test based on a random sample selected from all of the operators, the size of the sample being determined by use of probability sampling theory, 2) use of graphic quality-control charts on every machine showing ratio of output to input, and 3) plotting of total annual input (x axis) versus total annual output (y axis) for each of the past four or five years; then the point for the year in question would have fallen near the points for the other years, showing no substantial changes in productivity or output.

This example showed how a manager came to a false conclusion based on a biased sample. The problem was the lack of an adequate control record system to show how output was related to input. He needed the help of a professional statistician, not a detective or psychologist. The psychologist who broke the case was not using psychology as much as statistics; through him they discovered the wide variations in the ratio of output to input among machine operators. The operator could introduce variability in at least 20 different ways. The hypothesis that semiautomatic machinery led to a constant ratio of output to input was false. This illustrates that what management thinks is the real problem may not be the problem at all. It shows that operations should be based on sound information collected where applicable by modern statistical techniques. If anyone was to blame for the false inference, it was the manager himself, who didn't know how to manage.

9.   Human errors account for many fatal airplane accidents. Emphasis should be placed on prevention of errors, as illustrated by these examples:

- An airplane took off from National Airport, Washington, D.C., with ice on its wings. The plane rose about 300 feet and then crashed on the 14th Street bridge. Seventy people were killed. Cause: ice on wings.
- Flight 1713 took off from Stapleton Airport, Denver. It barely got off the ground when it flipped over. Twenty-eight people were killed. Causes: ice on the wings, inexperienced pilot, takeoff angle too great.
- The pilots of Flight 2268 took off from Detroit and neglected to set the wing flaps. As a result, 155 were killed, and only one passenger survived. Cause: oversight in checking all instruments and controls.

10.   The U.S. Department of Commerce had a sampling problem. The Department had contracted out a nationwide sample to a well-known firm. After several months, a team of three was asked for a report.

The Secretary wanted to get other opinions and asked that sampling experts from the National Bureau of Standards and the Interstate Commerce Commission attend a conference at which the report was to be made.

The contractors described a strange way of sampling: Select all the documents on the 17th day of the month. It was pointed out this was not a probability sample. This led to the cancellation of the contract for a nationwide sample of carload waybills.

11.   Medical-care errors are not uncommon. The following examples show what can be done to prevent them. Mix-ups are dangerous.

- A patient is given carbon dioxide, not oxygen. Institute clear identification; put oxygen in a separate room.
- A patient is given cocaine, not phenobarbital. Place drugs in separate cabinets and make clear identification.
- Two anesthesia machines have defective oxygen valves. Test before using.
- An oxygen tank is filled with argon. Make clear identification. Check deliveries.
- The Centers for Disease Control tested 3000 laboratories and found that 14 percent of the tests were in error. Needed: quality control over laboratory work, especially errors. Keep a time chart on a number of errors made by type of test. The Health Finance Agency sets a 5-percent error rate as acceptable under Medicare. A zero error rate is the ideal goal. How many died because of these errors?
- A survey in Arizona showed that:

  Some nursing homes in Arizona make medication errors 50 percent of the time.
  Thousands have died after getting wrong dosage.
  One of seven doses of medicine is in error. Put quality control on medicines and dosages by patients. A time line will identify each medicine and dosage.

- A nursing home in New York got two persons confused, according to a press report. The person who was alive was sent to a mortuary. Introduce tag identification.
- Two students were taken ill during lunch in an elementary school in Denver and were sent to a hospital. The cause of their illness was the way pizzas had been prepared. Cooking oil and oven cleaner were in lookalike bottles with similar labels. Furthermore, they were kept in the same place. . . . The pizzas had been prepared using oven cleaner! Many other students complained about burned throats and stomach pains. This is another case of error due to interchanging of containers.
- Needless surgery frequently occurs.[21] The Rand Corporation studied 5000 Medicare patients relative to four medical procedures used in surgery. The findings were:

  Procedure 1: 65 percent were done for inappropriate or questionable reasons

Procedure 2: 17 percent were clearly inappropriate
Procedure 3: 17 percent were clearly inappropriate
Procedure 4: 5 percent died due to this surgery

"Evidence exists that an ample number of physicians treating Medicare patients are performing inappropriate or unnecessary surgery. Some are charging too much for the procedures they perform, and some are guilty of both."

As a result of the survey, the U.S. Office of Management and Budget has suggested a reduction of payments to physicians for eight kinds of surgery.

12.   A postal employee says a 2-percent error rate is satisfactory—0.02 × 100 billion pieces of mail = 2 billion pieces in error. Not a record to boast about.

A Social Security official says that an error rate of 0.5 percent is excellent. Social Security issues 37 million checks each month; one-half of 1 percent is 185,000 errors each month. Not a very good record. These show the danger of a small percentage applied to a large base.

13.   Construction is not free of accidents. Deaths are caused by digging deep trenches that cave in on workers. Shoring up the sides of a trench would avoid this.

An overhead pass in Denver was about to be completed. When a final 60-ton block was placed, the west support collapsed. Faulty construction on the west pillar was the cause. One man was killed and another injured. The planners, the builder, and the Colorado Highway Department were involved. Construction of the west pillar was not finished.

14.   Railroad collisions: Three railroads dispatched trains on a single track:

1. In the Denver area, a freight train employee read "yesterday" as "today" on a posted notice. Consequently, he did not pull off onto a side track as the southbound train expected him to. The collision occurred right at the U.S. Highway 36 overpass. There would seem to be a safer way to handle the situation: two-way radio, red-green light, a red-green light in the place where he read the wrong date.
2. In Iowa, two trains collided but there was no explanation of what caused the collision.
3. A Conrail freight train left its side track and pulled directly in front of a speeding Amtrak passenger train. According to reports, the

Conrail engineer ran a red light and was guilty of other violations of safety. This alleged that he knowingly violated the rules of the road.

15.   Safety is the primary quality characteristic of such industries as health, transportation, construction, and nuclear power. Causes of disasters are:

- Ice on airplane wings
- Two-way rail traffic on a single track
- Lack of shoring on a construction job
- Mix-up of gases and medicines in a hospital
- Defective anesthesia machines in a hospital

Safety was a critical item in both of Three Mile Island (TMI) and the Chernobyl disasters. At both of these disasters, management and employees didn't know what to do in an emergency.

At TMI, the explosion did not melt the core, but workers did not know how to operate safety systems. They were not trained in nuclear physics, so they did not know the safety measures to take.

TMI was a small accident compared with Chernobyl. TMI's core was protected; Chernobyl's was not. The TMI process was stable; Chernobyl's was very tricky.

In both cases

- Employees did not understand. They worked against the safety system. They were to blame.
- Management was to blame.
- Employees were not trained in safety.
- Employees were not trained in nuclear physics.

16.   The wrong method is sometimes used.[22] Some years ago I visited the receiving department of a large manufacturing company, where I observed an inspector using a micrometer on a sample to test the conformity of the diameter of a cylindrical part. The sample plan, he explained, came from Military Standard 105, which contains a wide variety of sampling plans applicable to *attributes that are counted, not measured.* He could have reduced the sample size by about 50 percent and still obtained the same protection by using the standard deviation estimated from a sample. Furthermore, if the manufacturer put the diameter under statis-

tical control so that an estimate of the population standard deviation could be obtained, the sample could have been reduced another 50 percent. The sample was two to four times as large as needed to obtain the protection sought. This was pointed out to the inspector, and it was suggested that he buy Bowker and Goode's book *Sampling Inspection by Variables.* This shows the need to use the most efficient sampling techniques in a particular situation, and the danger of using, without adequate technical knowledge, compilations of sampling plans in particular and statistical formulas in general.

17.   Poor methods may be in use. The tour method is used to sample the work of employees or machine operations in a factory, office, or other work site. An observer walks through the factory or office, sometimes using a random time start, and tallies on a form whether each worker is idle or not and what task is being performed. He may tally whether a machine is idle or not and the reason. This procedure is called work sampling, and the following assumptions are made: 1) the number of observations is considered the size of the random sample; 2) the tallies or counts follow the heads-or-tails model (binomial model); 3) the various observations are independent; and 4) the presence of the observer does not significantly affect the working habits or performance of the employees. Every one of these assumptions is open to serious question. The physical operations performed simply cannot be matched with binomial distribution theory or with probability sampling theory. Instead, use random time sampling.

18.   A problem arose in measuring the weight of cigarettes being produced at the rate of over 2000 a minute. A large sample was selected and analyzed. This analysis took over two hours. When they returned to the production line, over 240,000 units had been produced! This was not quality control.

Twenty samples could have been drawn on the factory floor. With a hand-held calculator, $\bar{X}$ and R charts could have been produced in 10 minutes. This would allow 20,000 to get by. Automation requires *speed on the site,* not one or two hours' delay in a quality-control office. Even 10 minutes is too long. We need to concentrate on effective quality-control measures applied to automation.

19.   A canceled bank check can save a person from other's errors. Examples are:

- An insurance policy premium was paid. The insurance company threatened to cancel if payment wasn't received. Sending the company a copy of both sides of the canceled check settled the matter.

- Several books ordered from a book jobber and paid for. The books were sent in two shipments weeks apart. With each shipment the customer was billed for the complete order. Only a canceled check and a list itemizing each book title and its cost, to show that the check covered the *total bill*, finally settled the matter.
- A magazine gift subscription was paid by mail. Inquiry showed that the person had not received the magazine. Sending a copy of both sides of a canceled check finally settled the matter, and the person received the first copy of the magazine.
- A canceled check saved one year's payment of Social Security tax. Checking with the Englewood, Colorado, Social Security office led to the discovery of no credit for 1972. The taxpayer, a consultant, had paid the full amount. The following documents were submitted to correct the error: a 1972 tax return, a copy of an SE form, and a copy of a canceled check for all federal taxes. Either the IRS or Social Security, or both, made an error. The person was finally given full credit for payment of his 1972 Social Security. Only the canceled check led to the correction of the error.

20.   In this example of an effective method, a federal government agency needed to estimate an industry's dollar shipments and tons of carbon and alloy steel for the next quarter.[23] Aggregates and sample values for the past quarter were known. The population of 381 companies represented all the known companies in the industry, the sample size n was set at 50 companies, and 15 strata were used based on the past distribution of shipments, with the five largest companies being included 100 percent. Sample companies were selected at random from each stratum using a table of random numbers; the number of companies varied from two to eight per stratum. The allocation was close to optimal.

The estimate for the frame aggregate was a ratio estimate:

$$X = \sum_j \frac{Y_j}{y_j} x_j$$

where

$j = 1, 2, \ldots, 15$ strata
y_j = Known sample aggregate from past quarter
Y_j = Known population aggregate from past quarter
x_j = Sample aggregate for next quarter submitted by companies

The two sets of sample values were derived from the same sample of companies. This method of estimation, derived logically and tested experimentally by comparing it with the use of the simple ration N_j/n_j, was later called a ratio estimate requiring known data from the past.

For the fourth stratum, for example, the estimate was

$$X_4 = \frac{7589}{965}(1235) = 9,712$$

In $1,000s this was $9,712,000,

whereas the actual value was $9,589,000, giving a difference of 123,000 or 1.3 percent. For the aggregate of all 15 strata, $X = \$161,105,000$, and the actual obtained later by a 100-percent tabulation was $162,267,000, a difference of $1,163,000 or a deviation of 0.7 percent. The same method was applied to tons of carbon steel and alloy steel, with errors of 1.0 and 4.3 percent. This sample illustrated the improvement made by using many strata with relatively small random samples in each but with a cutoff of the largest companies and with an improved method of estimation (the ratio estimate) applied to each stratum. This was a breakthrough due to improvement in probability sampling.

21. A high-level Pentagon official stated: "We do sixteen million procurement actions a year. If you're better than Ivory Snow, ninety nine and one half percent pure, that's still one-half percent mistakes. That's 8000 a year!"[24]

This figure of 8000 is an error. The correct answer is 10 times as large:

$$\frac{1}{2} \times \frac{1}{100} \times 16,000,000 = 80,000.$$

Not an excellent record!

Another industry claims that Ivory's 99 44/100 percent is a very good standard to meet.

A Social Security official stated that an error rate of 0.5 percent was excellent performance on 37,000,000 checks issued monthly. This meant 185,000 checks were in error every month!

This is the danger of using small percentages applied to a large base.

Do it right the first time; there may not be another time. There may not be a second chance to do the job right. In many situations, the first time is the only time to avoid a dangerous or fatal error.

Correct knowledge and behavior are the keys to doing the job right the first time.

Previously noted examples (item 15) show critical situations in which knowledge, such as the following, can prevent fatal errors:

- Knowledge of when to de-ice the airplane
- Knowledge that the supplier is giving the hospital the right gas
- Correct identification of the contents of containers such as bottles, cans, and cylinders
- Knowledge of how a nuclear power plant's safety system works
- Knowledge of how centrifugal force acts relative to speed—the cause of many automobile and truck accidents

One error is all you can make. There is no rework, and there is no chance to correct the errors.

The Domino Effect of Errors

Making an error may be just the beginning of a long series of costly consequences.

Example:

1. A cylinder of carbon dioxide is mistaken for cylinder of oxygen.
2. A man dies.
3. The coroner rules that he died of natural causes.
4. The hospital starts training class on handling gases.
5. A son-in-law threatens to sue for negligence.
6. The hospital studies storage of gases.

Example:

1. An error made by a real estate broker has the buyer pay $314 a month in advance.
2. The buyer receives a computer bill for $314.
3. In a telephone call, he explains to mortgage company it is in error.
4. He receives a second computer bill for $314.
5. He makes another telephone call.
6. He calls the vice president; she promises to correct the error; the amortization schedule will cost $5.
7. There is no action; he receives a third computer bill.

8. He drives 55 miles to the company's office to correct the error.
9. He asks for a computer expert.
10. He finally finds a woman who knows the computer.
11. He explains the $314 error; asks for amortization schedule.
12. The woman goes to a computer station and corrects the error immediately; the amortization schedule will cost $2 and will be read out in two weeks; the vice president was mistaken and negligent.
13. The schedule arrives in two weeks.
14. It required months to correct this error.
15. The computer was not programmed for advance payments.

Example:

1. A customer is charged $166 for clothing she never bought.
2. She calls the store and reports the error.
3. She calls the central billing office and explains the error.
4. The central office claims that the customer "forgot" or "gave credit card to relatives."
5. The customer is again billed for $166.
6. The sales slip, dated on a Saturday, is found at Store 6.
7. The customer never bought at Store 6 on a Saturday, or at any other time.
8. The customer receives an affidavit that asserts her credit card was used illegally.
9. The manager of the store reports the billing error.
10. Central billing issues a new number to the customer.
11. The customer never receives any admission of error or an apology; the company remains silent. Was the company protecting a faulty job in Store 6 or the billing department?

11.   Reduce Delays and Lost Time

The major question for some service organizations to answer is: How many employees must be available to keep waiting times at an acceptable level? The question is how to plan service without making people wait beyond their "tolerated" time. Doctors make appointments every 15 minutes. This results in waiting 15 to 60 minutes, and people accept this. Other waiting times are shorter.

Waiting for work occurs in some service units. In a data-processing division, actual waiting time during one week of 40 hours amounted to:

Activity	Percent waiting for work
Graphic arts	9.0 percent
Typing and proofing	5.9 percent
Compiling	4.2 percent
Punched card file	7.2 percent

These data were collected by a carefully designed random time sampling. Steps were taken to reduce these figures to about 1 percent by better scheduling of the work. The purpose of the sample was *not* to control; however, the need for control, once the data were calculated, was obvious.

All the other 20 organization units had a waiting rate of 1 percent or less.

Wasted time covers a broad area. Some examples are:

- Making personal telephone calls on the job
- Coming late to work
- Taking a long lunch hour
- Being absent without leave
- Visiting; ignoring customers
- Taking long coffee breaks

It is easy to use *excessive time* to do a job such as an auto repair, a mail order, nose and throat treatment, an insurance claim, doctor's services, moving, delivery, or filling a book order by mail.

Idle time means doing something not required by the job. This includes

- Loafing
- Making personal telephone calls
- Socializing with other employees
- Taking long breaks
- Doing something personal
- Ignoring customers
- Waiting for work

Transportation time (shipping time) may be out of control for many reasons:

Buyer:

- Order is not clear; some key word omitted
- Delay in making request
- Does not use fast communication
- Delay in mailing
- Is not equipped for just-in-time process

Supplier:

- Temporarily out of stock
- Delay in shipping room
- Wrong address
- Delay in internal processing
- Does not use fast communication
- Does not use just-in-time process

Lost Time

Lost time is not due to an error that is corrected immediately. It is due to conflicts over errors, as reflected in the following:

1.   A businessman in Denver deposited $1842 in a large Denver bank. The bank recorded the deposit as $18.42. The customer called attention to this error and asked the bank to correct it.

The customer got a notice he was overdrawing his account. He responded that he did not make an error but the bank did, and he again pointed this out to them. Still he received notices that his account was overdrawn.

Now comes a new angle to this story. The customer sued the bank for taking his time from his business! The verdict: the amount on deposit was corrected, and the customer received damages for the time he lost from his business.

2.   In another case a judge fined a dentist for too much delay. The patient arrived five minutes early for his appointment but waited 77 minutes. He then walked out. The court awarded him $85.03 for waiting in the office and making an 84-mile round trip, and $30.30 in court costs. "The judge stated that when someone makes an appointment he is entitled to the time."[25]

3.   When a house buyer settled the purchase of a house with a real estate agency, the first month's payment of $314 was included. The mortgage company's computer was not programmed to handle an ad-

vance payment. The home buyer received notices and penalties, even though the $314 had already been paid.

Letters were written and telephone calls were made to the mortgage office, which was 55 miles away. Finally, a telephone call was made to a vice president; she promised to settle the problem. Nothing was done. Finally, the buyer drove to the mortgage building and asked for a computer operator. One was available, and when the problem was explained to her, she corrected the error in 30 seconds!

It required two letters, two telephone calls, a trip of 110 miles, an interview with a computer operator, and about three months' total time before this error was corrected. Correcting this error cost the customer about $200.

4. This case involves bringing an amortization schedule of a house up to date when advanced payments were made.

Under a 20-year mortgage, monthly payments are itemized for 240 months. A person is allowed to make advance payments not to exceed 20 percent of the total due.

A home buyer wanted to make a payment of about $4000. To do this, one starts with the last total amount owed, as shown on the amortization schedule:

Last total amount	$24,601
Payment	−4,290
Payment on new amount	$20,311

(Shifting from the last total amount to $20,311 allows the buyer to skip 12 monthly payments.)

The problem was to submit a payment so it matched the amortization schedule. This was done. A letter explained the calculation. A woman at the company was called to explain what was done. When a second call was made, a man answered. He couldn't understand the arithmetic or the amount to be paid. It had to be less than 20 percent of the amount due.

Finally, after prolonged long-distance telephone conversations and about two months, the amortization schedule was updated correctly.

In the last two cases, the customer made telephone calls, wrote letters, and traveled over a hundred miles to correct an error the company made. This was lost time, and the customer did not even receive a "thank you" for it. The cost of correcting this second error was about $150.

Delay

The person who wants service today must get used to waiting. At the supermarket, bank, post office, insurance office, drugstore, tax office, doctor's office, gas station, hotel, or restaurant, waiting is what one accepts as part of life. Waiting, however, has its limits.

The quality of service depends upon the employee waiting on the customer. This service is discussed later. Most people accept a moderate waiting time. An old-time newspaper man says he will wait in a doctor's office for 20 minutes and no more. If the doctor doesn't get to him in that time, he leaves.

Here are examples of waiting times from a recent Health and Human Services Department study:

> waiting for a doctor's appointment: 7 days
> waiting in doctor's outer office: 29 minutes
> waiting in emergency room: 38 minutes
> waiting as hospital outpatient: 45 minutes

These are *average* waiting times; some wait a shorter time, others wait a longer time. This waiting time causes inconvenience and lost time at work, and can be serious for those who need immediate attention.

Delay Versus Convenience

During a six-hour test, two staff writers from the *Rocky Mountain News* using stopwatches compared waiting times at counter service versus drive-in service, and waiting times at regular lines versus express lines (Table 4.1).[26]

The writers spent a total of 50 minutes, 49 seconds in drive-throughs, compared with 41 minutes, 22 seconds when they went inside. They made 22 stops: 10 fast-food restaurants, four banks, six supermarkets and one stop each at Target and K-Mart.

According to this study you spend more time using the convenient drive-through. The table shows that you would save time at three banks and at three fast-food restaurants by being served at the counter: According to this test, then, you should avoid drive-throughs if you want to save time.

Table 4.1
Counter Service Versus Drive-Throughs

		Time waiting (min)	
Time of day	Place	Counter	Drive-through
9:40 A.M.	Colo. National Bank	0:50	1:00
9:55 A.M.	1st Int'state Bank	1:18	2:35
11:56 A.M.	McDonald's	1:08	6:37
12:08 P.M.	Wendy's	4:17	4:07
12:16 P.M.	Burger King	6:35	4:05
12:34 P.M.	Taco Bell	8:50	10:02
1 P.M.	Arby's	3:36	7:24
2:49 P.M.	Capitol Federal	1:17	2:15

Source: Ref. 26.

Avoiding Delay by Choosing Express Lanes

Sometimes the regular supermarket lane seems to be moving faster than an express lane, but in eight comparisons of express and regular lanes, regular lanes came out first only once.[26] The study showed that the problem of delays is solved by using the express lane (Table 4.2).

The writers spent a total of 43 minutes, 13 seconds in regular lanes, compared with 31 minutes, 16 seconds in express lanes.

Delays in a regular lane are caused by:

- Writing checks
- Buying lottery tickets
- Buying food stamps
- Getting money, or wallet, from handbag
- Getting someone to help carry groceries to car

UPC Errors and Delay

UPC stands for the Universal Product Code, which is used in supermarket computer systems to itemize and total a sales ticket at the checkout.

The computer is laser-operated, and the prices are programmed into it. In one chain store, prices are programmed every Wednesday.

If prices are changed, as for a sale, the computer price may not be the

Table 4.2

Regular Lines Versus Express Lines

| Time of day | Place | Time waiting (min) | |
		Regular	Express
10:55 A.M.	King Soopers	7:39	6:45
11:10 A.M.	Safeway	3:49	1:30
11:35 A.M.	King Soopers	7:16	2:56
1:30 P.M.	King Soopers	11:23	6:30
2:10 P.M.	K-Mart	2:41	1:59
2:25 P.M.	King Soopers	4:44	4:51
3 P.M.	Safeway	2:10	0:45
3:30 P.M.	Target	4:51	1:55

Source: Ref. 8.

same as the posted price. This may lead to overpricing. A New York study showed that, on the average, about 5 percent of the items were mispriced in favor of the supermarket. A Colorado study showed a similar trend.

This means the prices should be checked to avoid errors and delay. One has to beware of shopping on Wednesday or any other day when new prices are programmed.

Human errors are no excuse for this error rate. Prices should be changed on the item *before* they are programmed so the correct price can be entered in the computer. Zero error is the goal.

12. Ensure Safety

There are many kinds of unsafe situations.

- Health services: Protect the patient from gases and drugs.
- Transportation: Protect the passenger, equipment, and the driver— pilot, truck driver, bus driver, locomotive engineer.
- Factories: Protect the worker and machinery.
- Construction: Protect the workers, equipment, and construction materials.
- Power plants: Protect the workers and equipment.
- Household: Protect from fire, gas, and electrical hazards.

Safety Program

Creating safe products is one program; creating safe operations is another. The first ensures a safe product; for example, creating adequate tensile strength. This requires total quality control.

Operating safety involves human error. It requires long training, experience, and attention to details to avoid these errors.

A human error program is a continuous program for discussing all kinds of errors and how to prevent them. Critical errors are given top priority, followed by errors that are less critical but just as important to prevent.

The Aircraft Owners and Pilots Association conducted a nationwide telephone poll of 1000 adults regarding what should be done to increase air travel safety. The results were:[27]

	Percent
Improve plane maintenance	44
Improve air traffic control system	20
Improve airport capacity	11
Shift airline flight schedules	9
Limit private aircraft at major airports	7
Modernize weather-detection systems	3
Other	6
Total	100

Airplane maintenance heads the list. Other major factors omitted from the list are the training, experience, and flying habits of the pilots.

Example: It was possible that nearly half of a shipment of transit buses would have to be recalled because of defective bolts used in a rear axle.[28] The Denver transit company was notified of possible problems after a Los Angeles transportation agency complained the bolts had not been properly tested. About 300 of 650 buses were equipped with the suspect bolts. The buses were built by a Colorado firm. Quality control was needed (Deming's point 4).

Several types of service organizations particularly need a safety program:

• Transportation
• Health services
• Construction

- Chemical plants
- Nuclear power plants

This program should include instruction in the following:

- Places where danger can arise
- Steps to take to minimize danger
- What to do if a dangerous situation occurs
- Care to be taken to avoid a dangerous situation

Lifesaving methods are to be used. Care should be taken to patrol dangerous areas to prevent any accident from occurring.

We have already given several examples of violation of safety rules. In only one case were any steps taken to prevent these errors.

Examples of Unsafe Conditions

We discuss several ways of moving toward quality.

Airline Pilots

Safety depends to a great extent on the health of the pilots; fatigue and workload are major factors. Pilots are often subject to:

- Crossing six to eight time zones
- Night flying
- The crucial maneuver of setting the flaps
- Long hours, long flights
- Distractions in the cockpit

The solution is fewer work hours and longer rest periods. Closer attention must be paid to every detail when preparing for a takeoff or landing.

Deregulation leads to lower fares; one way airlines offset this is to make pilots work more hours, but this is fatiguing. It also reduces maintenance, which increases the danger of an accident.

On longer transoceanic flights, arrangements might be made for a relief crew.

A quality-control system is needed in the operation division as well as in the maintenance department.

Food Handlers and Food Management

Colorado state auditors warned state health officials that they should do a better job of recording the inspection of 13,000 establishments, including restaurants, food carts, convenience stores, and school lunch programs.

Both business owners and the FDA agree that state and local inspections lack uniformity. Inspections are required every six months but are usually carried out about every 300 days.

Problems arise in critical areas:

- Food storage
- Water temperatures
- Disinfectant levels
- Need for inspection in all seasons
- Contaminated water sources
- Storing cleaners in food containers
- Rodents
- Personal cleanliness
- Lack of sanitation

No adequate records are kept of state or local inspectors' visits. What is needed:

- A quality-control system covering the entire food service, not just what the state health inspectors deem important
- More inspectors to prevent epidemics and disasters
- More inspections

Nursing Homes

Quality is a rare characteristic in nursing homes. The following conditions were found in 184 Medicaid-certified nursing homes in Colorado:

- Twenty percent did not investigate life-threatening situations, such as an overdose of medicine, within 24 hours.
- Thirty-four percent did not investigate serious complaints within 72 working hours (72 hours is the prescribed number).
- The federal ombudsman handling complaints in the state on such issues as rates and stolen personal property received about 1200 complaints in 1987.

- Every home violated at least one Medicaid requirement in 1987.
- Seventy-eight percent repeated less serious violations within the last four years. "Less serious violations" includes such actions as failing to keep patients clean! This is not quality control.

Quality control can be applied to:

- Life-threatening situations
- Complaints
- "Less serious violations"

Complaints include theft of personal items, administration of wrong medicines, theft of residents' money, poor meals, abusive treatment, neglect of personal care (cleanliness), lack of privacy, and unanswered calls.

Factories

L. D. Stimely of the Lapp Insulation Company described how the company used nurses' records and personal interviews to reduce workers' injuries.[29]

> Interviewing the injured people or reading the [nurses'] injury records yielded a Pareto Diagram of use in determining what activity caused the injury. Surprising in its magnitude was 'lifting,' which was responsible for about two-thirds of the back injuries . . . the youngest group, the 18–22 year olds, was the highest.

This countered the assumption that the oldest group should have the back injuries.
The causes were as follows:

	Percent
Lifting	63.3
Unknown	13.3
Non–work-related	10.0
Falling	3.3
Twisting	3.3
Body motion	3.3
Total	99.8

The major reason this study was undertaken was to provide management with the information it would need to take corrective action. This objective was accomplished quite well and there followed an abundance of activity in safety areas. These are summarized below:

1.   Injuries from heavy lifting: Cranes and hoists were installed where necessary. Job classifications and MOST sheets were reviewed for excess physical requirements or unrealistic standards. Job studies were made and lifting standards revised.

2.   Injuries from working in unfamiliar areas: All maintenance procedures were revised to show safety requirement, precautions, and necessary safety equipment. It is required that all maintenance personnel review the procedure prior to performing the task.

3.   Finger, hand, feet, and eye injuries: Safety equipment is required to be worn and includes gloves, safety shoes, and safety glasses.

4.   Back injuries: Besides those items shown under heavy lifting, the nurse showed movies and gave pamphlets on proper lifting techniques to all employees. The youngest group of employees were given special instructions. They were primarily being injured because they felt that they were too healthy to be hurt and were taking short-cuts when it came to lifting.

5.   Inattentive work: A meeting was held with all our employees to discuss the results of this study. Motivational programs were tried and safety contests were held.

6.   Future problems: We hope to head them off before they occur by hiring a safety engineer, and contracting an outside machinery and automation company to review the safety aspects of all of our equipment.

Schoolbuses

A Boulder School District bus carrying 40 sixth-grade students home from a mountain field trip rolled off Colorado Route 7, landing a few feet from South St. Vrain Creek. One student was killed and 21 were injured.[30]

The teacher who was driving had complained about the brakes earlier when the bus was 1500 to 2000 feet above the site of the accident. She had taken lessons in mountain driving. It was therefore thought that the brakes had failed, and the assistant superintendent suggested installing auxiliary brakes in all buses used in mountain driving. These might not have prevented the accident.

On some mountains in the Rockies, easily readable signs are posted:

STEEP GRADE
SHIFT TO LOW GEAR

No one has yet suggested that braking action is greatly increased by shifting down to the lowest gear. This could have been done on the schoolbus.

I have had two experiences in which brakes started to burn out. One occurred while driving over Logan Pass on Going to the Sun Highway in Glacier National Park in Montana. The altitude of Logan Pass is 6000 feet. About a quarter of the way down, we smelled the scorched brakes. We turned off at a nearby turnout spot to let the brakes cool off. Then we went the rest of the way in low gear—at about 10 mph. This car was a Country Sedan with a 225-h.p. Thunderbird engine. Downhill speed on a long stretch can easily burn out the brakes.

On another occasion we were caught in the clouds at the top of the Trail Ridge Road in Colorado. At an altitude of 12,000 feet, this is the highest road in the country. The clouds extended down to the 8000-foot level. We shifted the transmission to low gear, which slowed us to 10–15 mph. Road visibility was only about 100 feet, just enough to follow the winding road.

In the Colorado schoolbus accident, three of four airbrakes were out of proper adjustment. This shifted all the braking action onto the fourth brake, which would cause the vehicle to pull to one side of the road.[31]

This raises a serious question about the maintenance of these buses. Quality control includes maintenance.

The school district plans to install an auxiliary set of brakes on mountain buses. This raises a question: Will not a second set of brakes burn out just as the first set will on these miles of steep grades?

13.   Avoid Purchasing Defective Products

All service organizations purchase defective products. The problem is how to prevent defects from occurring.

1.   An urban rapid-transit company placed an offer for over 100 buses to the *lowest bidder*. Of the first 30 buses delivered, 15 were found defective. They were *not* returned to the factory for correction of defects. They were repaired in the company's garage. Quality control would call

for the prevention of these defects in the factory. The transit company approved defective machinery (Deming's point 4).

2.   We have described how one hospital used carbon dioxide instead of oxygen, another used argon instead of oxygen, and still another used two defective anesthesia machines.

3.   The U.S. Food and Drug Administration found that some pacemakers manufactured both in this country and abroad were defective because of poor design and assembly. The FDA warned all users but could not control quality at the source. All it could do was cite the manufacturers for producing defective products.

The FDA found a manufacturer that mixed penicillin and aspirin. The manufacturer stopped distribution. We don't know what steps were taken to separate these lookalike drugs.

The FDA has discovered the following defects in products:

- Labels switched on bottles of two drugs
- Wrong label used; misbranded
- Wrong count of pills in bottles
- A transposition label error in which 8.3 mg was used instead of 3.8 mg
- Can seams were not sealed

4.   We have pointed out how a city government may receive defective products when it buys water meters for household use. In one instance, quite a number were unreliable or would not give accurate readings on an auxiliary meter installed at a convenient location on the house.

Rain and lawn water short-circuited the meter on at least one house. Also, the meter in the ground did not work. It was replaced by another meter. Finally, after about three years, the utility got the system to work.

The manufacturers had never heard of quality control; they were not even making a product that worked.

One has to verify data and operation of purchased products to ascertain if they are the correct products. Manufacturers are obligated to produce an acceptable-quality product—a product that works.

5.   Calendars purchased at a retail store required sequins to be sewn on. Each calendar had a large number of sequins of different colors to be sewn in the proper places. The colors were not placed in separate packages. Even worse, sequins of one color did not have any holes punched in

them! They could not be used. This is an example of lack of quality control.

6. Inspection by the purchasing company is a waste of time and money. The manufacturer should submit proof of quality control—proof that the product is free of defects and meets the buyer's specifications.

Inspection assumes defects and errors. It assumes quality has *not* been built into the product or service. The goal: Get rid of inspection as much as possible.

7. The U.S. Customs Service seized 87 tons of suspected defective bolts during raids on 11 Western bolt manufacturers.[32] The defects were in the bolts' tensile and shearing strengths. An agent claimed that the bolts contained too much boron and not enough carbon.

Bridge failures on the East Coast and tank failures at Fort Carson in Colorado and other military facilities were the result of substandard bolts. "Just about every major user of fasteners probably has an inventory of counterfeited and mismarked bolts," an agent said. "That means the public has been riding on planes and buses that are using these defective bolts." He added that defective bolts are much more apt to break down and shear under torque.

The bolts were manufactured in the Orient but were passed off as American-made.

Where is sample testing for tensile strength and shearing strength? Why weren't the bolts tested to meet U.S. standards? There is not a single piece of evidence of quality control by users in the United States! They settled for defectives.

8. During the 1980s, HUD (Department of Housing and Urban Development) lacked an effective fiscal accounting system, resulting in losses of hundreds of millions of dollars.[33] Denver is trying to find out what happened to $7.2 million.

A woman embezzled $5.5 million. She said she gave part of it to the poor and spent the remainder on herself.

An agent in Colorado embezzled $1 million.

An Alabama broker kept more than $2 million. He spent nearly $180,000 on cars.

HUD's rule is that as soon as a piece of property is sold, the money is immediately sent to HUD in Washington. This is the way the above transactions should have been handled.

The total loss to HUD is estimated to be about $2.5 billion due to favoritism, patronizing friends, embezzlement, and lack of accounting.

9. The General Accounting Office (GAO) reviewed the accounting

practices of 11 federal agencies during the last three and one-half years (1977–1980).[34]

About $218 million in accounts receivable at the Labor Department had not been recorded as such. Thus, no efforts were made to collect the money.

The Naval Fleet Finance Center failed to preaudit travel vouchers, resulting in $700,000 in overpayments in two months.

14. Take into Account the Learning Curve

There are three kinds of learning curves: the error curve (Figure 4.1), the production curve (Figure 4.2), and the unit cost curve (Figure 4.3). All three contribute to productivity.

Error curves reflect certain properties.

- They are irregular in shape.
- They may have plateaus.
- At time near 0, errors are many and large with little production.
- As time increases, errors tend to decrease, production increases, and unit cost decreases.
- The limit to errors is zero.
- No study has been made to show how close errors approach zero in practice.

Plotted error curves are irregular. At more than one interval, the curve appears horizontal, as though this is "the best we can do." These are

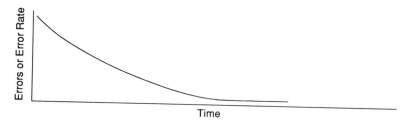

Figure 4.1
The error curve.

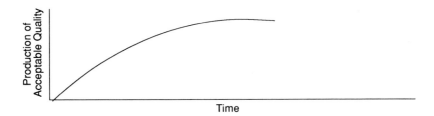

Figure 4.2
The production curve.

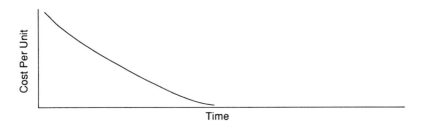

Figure 4.3
The unit cost curve.

simply plateaus with data that sometimes are considered in statistical control when they are not. An error curve may have two, three, or more plateaus. In a learning curve for data processing, three plateaus were found on a production curve.

In a learning situation, it is better to plot individual points to establish the actual situation.

The data required are:

- Number of errors in a work unit per employee per day
- Amount of quality production per employee per day
- Cost per unit equals average salary cost per day's work

Data are collected by the random time sampling method already described.

15.   Institute a Suggestion System

The suggestion system has had a mixed past. Most systems failed for several reasons; the major cause was management's neglect and indifference. The following changes are recommended:

- Suggestions should be collected weekly.
- Management should discuss them with the employees involved.
- There should be individual rewards for acceptable suggestions.
- Rewards should be based on amount of savings to the company.
- This should be an ongoing program.
- All acceptable suggestions should be implemented at once.
- Management should recognize each contribution before all employees.

Companies profit from employees' suggestions for quality improvement. Companies using this system receive thousands of these suggestions every year and reduce costs by hundreds of thousands of dollars.

The foregoing changes should be company policy. Management should announce leadership and continuous support of this policy.

Employees should be encouraged to take part in the system and submit ideas for improvements. Under the right approach, employees will submit hundreds, if not thousands, of ideas for quality improvements.

Example of a Suggestion System That Failed

A professional employee—a mathematical statistician—observed several places in which probability sampling could be used to an advantage. He suggested that whenever a sampling study be made, the proposed sample be reviewed by a sampling staff of three to ensure that quality data were being collected.

The suggestion, outlined in some detail, was dropped in the suggestion box. The employee was never contacted in person, nor was there any response for several weeks. A top official answered in writing that there wasn't any need for this change, that it would not improve the situation, and that it would mess up operations. This official would be affected by this change and presumably preferred the status quo.

A Successful System

Milliken and Company won the 1989 Baldrige National Quality Award for high-quality products.[35,36] Its program featured the following:

- Asked all employees to make quality-improvement suggestions.
- In 1988, employees made 115,000 suggestions, of which 85 percent were implemented.
- Organized 1600 teams to tackle manufacturing problems and 500 teams to work with customers.
- Gave workers the power to shut down a defective process or machine.
- In 1988, spent about $1300 per employee on quality-improvement training.

16. Understand How Quality Improves Productivity

Quality improves productivity in services in the following ways:

- By eliminating *errors*, quality yields hours to be used elsewhere.
- By eliminating *defects* in purchased products, quality yields extra time.
- By reducing time to *perform* an activity, quality yields extra time.
- By promoting *safety* and reducing unsafe conditions, quality yields an annual savings.

These improvements put more workers on the production line. This decreases wasted time, reduces costs, and increases productivity.

Improved productivity means the same work force does more work daily or weekly because workers perform more shrewdly using quality control methods. More production by the same workers increases productivity.

If you reduce a 6-percent error rate to 2 percent, you are adding 4 percent to productivity by working more shrewdly. This is due to knowledge, not to any new machinery or equipment, nor to any pep talks.

This is nothing new. *Industrial Quality Control* has carried articles that showed that quality control resulted in gains in productivity by shifting inspectors to production. What Deming did was to collect all of these components into a plan and show how they worked together.

How to Improve Productivity by Industries

Insurance

- Eliminating unnecessary paper handling

Banking

- Eliminating errors in checking accounts

Health

- Avoiding drug and gas errors

Retail trade

- Avoiding a billing error that creates problems for the company and the customer

Transportation

- Wasting time repairing newly purchased defective buses—send them back to the factory (immediately)
- Tools not readily accessible

Government

- Using a probability sampling instead of 100-percent coverage

Public utilities

- Avoiding defective water meters
- Helping customers avoid anonymous and wrong-number telephone calls
- Saving time by demonstrating how to use the telephone
- Eliminating waiting for calling party to identify himself
- Avoiding billing errors

Personal service, transportation

- Creating error-free reservations
- Using suitable billing form

Examples of Increased Productivity

- Reduce receiving inspection of purchased products; shift inspectors to production.
- Reduce safety inspection if doing so will not impair safety.
- Prevent errors so more can be accomplished daily, weekly, or monthly; use the learning curve.
- Avoid idle time waiting for work.
- Avoid idle time waiting for tools or repair parts.
- Develop a correct billing operation.
- Develop an improved probability sample survey.
- Substitute probability sampling for 100-percent coverage.
- Repair more things in the same amount of time.
- Shift some verifiers to production.

The Road to Productivity

The federal government issues studies on productivity in services:

- They are global.
- They consist of statistics.
- They do not explain how to improve productivity.

Quality control provides a sure-fire method that every organization can use to improve quality and increase productivity. The road to productivity is quality control.

Quality control involves knowledge. So does probability sampling. There is no need to invest in expensive machinery, equipment, laboratory apparatus, or computers. All that is needed is correct use of certain kinds of technical knowledge, including the science of statistics and the learning curve.

Selling Price Has No Relation to Quality

Some articles of quality cost more; others cost less. What factors influence quality?

- Clothing buttons are not sewn on to stay. This has been an age-old defect. There has been no improvement in eliminating this defect. Price makes no difference.
- A quality-made foreign automobile doesn't cost any more than a domestic automobile despite quotas and agreements.
- "One size fits all" for women's and men's clothing is very unrealistic. It is a retreat from quality improvement.
- Most houses and apartments are not built to satisfy the customer.
- An airline could not care less about 1) how a customer gets to the airport or 2) how a passenger gets home after a flight. No accommodation is made to get passengers from origin to destination.

This raises the question of how much the customer gains from an organization's quality-improvement program. The customer gets a better product but not a reduced price. In this regard, only the company gains.

In the example cited by Deming, a reduction of 0.1 pound in producing carbonless paper saves $100,000. A quality-improvement program reduced poundage from 3.6 to 1.3, saving $2,300,000. How does this gain affect the customer? Lower prices? No change? Higher prices?

Does the customer notice the difference? Probably not. The example emphasizes only the monetary gain of the company.

Customers Don't Buy from CEOs

The effectiveness of CEOs in maintaining quality depends on how carefully they pass on their commitment to all of the lower levels. They need acceptance of quality all the way down to the lowest-paid worker. Unless and until this happens, there is no company-wide quality program.

The CEOs of Ford Motor Company and General Motors have committed themselves to quality and made this quite clear in statements in *Quality Progress*.

But it takes a long time to accept the CEO's policies.

"Quality is job 1" is Ford's creed. Apparently Ford is making real progress in quality improvement. Whether this gain justifies a price increase is another matter. The Econoline bus has a tendency to catch fire. Six companies have filed a class action suit in a Denver court for damages incurred when their vans caught fire. In 1989, Ford recalled 1,950,000 cars in several models from 1984 to 1989 for safety-related problems involving engine fires, wheels running off, and emissions. Recent complaints concern peeling paint. Where has quality gone? The

CEOs' words are not enough. They have to get their quality commitment accepted and practiced throughout the organization.

Nor is General Motors any better. It is recalling 1,700,000 vehicles that have a cruise-control defect that could cause throttles to stick.

These recalls will cost millions of dollars. Apparently, to them, quality does not mean doing it right the first time. It does not mean zero defects. It does not mean conformance to requirements. It does not even mean fitness for use. The CEO's plans, procedures, and policies are not passed on to the working level.

Customers don't buy from the CEOs. They buy or are serviced by individuals who do the work:

- Salespersons
- Mail clerks
- Bank tellers
- Insurance agents
- Ticket agents
- Flight attendants
- Hotel and motel clerks
- Waiters and waitresses
- Tax clerks
- Utility billing clerks
- Telephone clerks
- Checkers
- Travel agents
- Repair persons
- Receptionists
- Factory workers
- Medical assistants

These people determine the quality of service rendered. Some follow the CEOs, many do not, but there is a trend to treat the customer in a civilized manner. Examples are:

- Salesperson—friendly, knows stock
- Mail clerk—knows domestic and international procedures
- Bank teller—knows how to handle deposits and withdrawals

Customers buy medical services from a physician, but assistants do most of the tests and measurements:

- Measurements: temperature, pulse, blood pressure, EKG, X-rays
- Blood sample
- Medical history
- Eye test

17.   Aim for Quality Improvement

The purpose of quality planning and quality control is the improvement of quality. Always the objective is quality improvement.

Quality improvements center on the solution of problems. Problems are selected so that a solution will result in a big gain, a breakthrough.

One common problem is a situation fraught with errors and defects that can be driven to zero, or close to it. Where one error can be fatal, a continuous safety training program is necessary. This applies to such services as the following:

- Airline services—ice on wings, faulty engine, metal failure
- Medical services—fatal gases, mixup of drugs, wrong drugs
- Other transportation—failure of buses, trucks, trains
- Construction—shoring, faulty design, construction work

Another problem is wasted time, including all kinds of unwarranted delays. Employees are responsible for some of this wasted time, management practices account for some more, and the system accounts for the remainder. Introduction of just-in-time saves time and inventory costs both in services and in manufacturing.

Friendly employees help bring back customers. To be successful, business must have repeat customers. Polite and courteous employees help a business do this. A business should not overlook the behavior of its employees in a quality-improvement program. Courses in quality behavior should be given to new employees, as well as to current employees, to emphasize the *importance of behavior* in determining quality.

18.   Understand the Cost of Non-Quality

The cost of non-quality is enormous but unknown.

Human Waiting

Profits are made in a doctor's office by overbooking and making patients wait 30 minutes to 1 hour over the appointed time.

Patients are kept in a hospital days after they should be discharged. Unnecessary surgery is performed.

Some supermarkets maintain short lines; others do not.

Millions of dollars are wasted because of excessive amounts of human waiting.

Lost Time

The cost of waiting for work, wasted time, idle time for those at work, lost time of customers due to errors, and various forms of delay can be enormous. This does not include the time lost by millions of unemployed workers.

Errors

Human errors cause disasters in hospitals, nursing homes, domestic air travel, trains, ships, nuclear power plants, and construction. The cost is incalculable.

Human errors cause a multitude of inaccuracies of lesser importance in all service companies. The errors include billing, computer operations, inefficient sampling, the danger of a small percentage, faulty laboratory work, and errors in paperwork.

The cost of these errors is hard to calculate but runs into billions of dollars.

Customers

Customers are faced with delays, errors, defective products, lost time, time occupied with unnecessary actions, idleness, and unwarranted be-havior and attitudes of the work force.

All these spell trouble and extra cost. All these unnecessary costs are borne by the customer. He has three choices: Accept them, complain, or shift to a competitor if one exists.

In transportation and health services, error situations can be fatal to some customers—passengers and patients. These costs cannot be calcu-lated.

Pleased customers are the only ones the company can count on. Dissatisfied customers are risky. The losses due to lost customers are magnified, since they spread the word of the poor-quality service and products they are receiving. This means an unknown number of poten-tial customers are lost.

Shoplifting and Employee Theft

The customer pays for theft in higher prices, or, if these items are taken by the company as tax deductions, taxpayers bear the burden. If both actions are taken, the customers get a double dose of penalties.

Employee Behavior

Poor-quality employee behavior creates a substantial but unknown cost due to lost customers. This behavior takes the following forms:

- Ignoring customers
- Making customers wait until some store-related job is completed
- Visiting with other employees
- Waiting on a customer out of turn
- Treating customers rudely
- Making a customer wait when a transaction is interrupted by a telephone call
- Charging a customer unjustly for a company error—employees play a major role in company errors

Much of this type of behavior is due to the company's laxness and general mismanagement.

References

1. Robert S. Kahn, Advanced accounting for advanced technological environments. *Science* 245:820, Aug. 25, 1989.
2. Charles Wiman. *Industrial Quality Control*, January 1949.
3. J. W. McNairy. *Industrial Quality Control*, July 1949.
4. Dale W. Lobsinger. *Industrial Quality Control*, May 1950.
5. J. M. Ballowe. *Industrial Quality Control*, January 1946.
6. R. H. Noel and M. A. Brombaugh. *Industrial Quality Control*, September 1950.
7. A. C. Rosander, H. E. Guterman, and A. J. McKeon. *Journal of the American Statistical Association* 53:382–397, June 1959.
8. A. C. Rosander. *Applications of Quality Control in the Service Industries*, New York: Marcel Dekker and Milwaukee: ASQC Quality Press, 1985.
9. A. C. Rosander. *The Quest for Quality in Services*, Milwaukee: ASQC Quality Press and White Plains, NY: Quality Resources, 1989, pp. 293–299.

10. J. M. Juran. *Quality Progress,* Aug. 1986, pp. 19–24. This is Dr. Juran's trilogy.
11. Wade Weaver. *Industrial Quality Control,* Sept. 1948.
12. Loveland, Colorado, *Reporter-Herald,* June 23–24, 1989, p. 4.
13. A. C. Rosander. *Case Studies in Sample Design,* New York: Marcel Dekker, 1977.
14. R. E. Heiland and W. J. Richardson. *Work Sampling,* New York: McGraw-Hill, 1957.
15. Gerald J. Hahn, W. J. Youden memorial address, 31st Annual Fall Technical Conference, 1987. *Statistics Division Newsletter* 9(1):8, Fall 1988.
16. Stanley Payne. *The Art of Asking Questions,* Princeton, NJ: Princeton University Press, 1951.
17. A. C. Rosander. *Applications of Quality Control,* p. 66.
18. *Rocky Mountain News,* Feb. 28, 1984, p. 1.
19. *Rocky Mountain News,* Dec. 15, 1988, p. 61.
20. A. C. Rosander. *Case Studies,* p. 144.
21. Surgery: Unnecessary? Inappropriate? Overpriced? *Parade,* May 21, 1989, p. 19.
22. A. C. Rosander. *Case Studies,* p. 143.
23. A. C. Rosander. *Case Studies,* p. 149.
24. Hendrick Smith. *The Power Game,* New York: Ballantine Books, 1988, p. 163.
25. *Rocky Mountain News,* Feb. 14, 1982, p. 2.
26. *Rocky Mountain News,* Nov. 21, 1989, p. 29.
27. *USA Today,* Aug. 10, 1988, p. 6A.
28. *Rocky Mountain News,* March 11, 1981, p. 6.
29. L. D. Stimely, Statistical thinking and management. Proceedings of the 44th Annual Midwest Quality Conference, Fort Collins, Colorado, Oct. 1989, pp. 72–75.
30. *Rocky Mountain News,* June 23, 1989.
31. *Rocky Mountain News,* July 5, 1989, p. 6.
32. Loveland, Colorado, *Reporter-Herald,* AP dispatch, July 1–2, 1989.
33. *Rocky Mountain News,* July 1, 1989.
34. Loveland, Colorado, *Reporter-Herald,* AP dispatch, Sept. 1, 1980.
35. *USA Today,* Nov. 3, 1989, p. 10B.
36. *USA Today,* Dec. 26, 1989, p. 5B.

Index

Accuracy, 67
Airline pilots, unsafe working conditions for, 121
Alden's (mail-order house), 18
Ambition, barrier between departments due to, 32
American Society for Quality Control, 1
Analysis of quality control data, 61
Anti-quality behavior and attitudes, 95–96, 138
  training salespersons to avoid, 27–28
Assignable causes, 7
Attendants, training for, 66–68
Attitudes (*see also* Anti-quality behavior and attitudes):
  differences of, 32
  of employees, surveying, 54–55

Automation, failure of, 61–62

Ballowe, J. M., 18
Barriers between departments:
  breaking down conflicts causing, 55
  elimination of, 31–35
    common causes, 33–35
    by instruction, 31–32
    resolving emotional differences 32–33
    special causes, 33
    types of barriers, 31
Behavior (*see also* Anti-quality behavior and attitudes):
  of employees, surveying, 54–55
Binomial count np, 68, 70
Binomial proportion p, 68, 70–71